W9-CSP-065

Analytical and Hybrid Methods in the Theory of Slot-Hole Coupling of Electrodynamic Volumes

Mikhail V. Nesterenko · Victor A. Katrich
Yuriy M. Penkin · Sergey L. Berdnik

Analytical and Hybrid Methods in the Theory of Slot-Hole Coupling of Electrodynamic Volumes

 Springer

Mikhail V. Nesterenko
Department of Radiophysics
V.N. Karazin Kharkov National University
Kharkov 61077, Ukraine
mikhail.v.nesterenko@univer.kharkov.ua

Victor A. Katrich
Department of Radiophysics
V.N. Karazin Kharkov National University
Kharkov 61077, Ukraine
victor.a.katrich@univer.kharkov.ua

Yuriy M. Penkin
Department of Information Technology
National Pharmaceutical University
Kharkov 61002, Ukraine
kit@ukrfa.kharkov.ua

Sergey L. Berdnik
Department of Radiophysics
V.N. Karazin Kharkov National University
Kharkov 61077, Ukraine
sergey.l.berdnik@univer.kharkov.ua

ISBN: 978-0-387-76360-6   e-ISBN: 978-0-387-76362-0
DOI: 10.1007/978-0-387-76362-0

Library of Congress Control Number: 2008921454

Printed on acid-free paper

9 8 7 6 5 4 3 2 1

springer.com

It is strange but true; for truth is always strange, stranger than fiction.

Byron

To
Ekaterina N. Nesterenko
and Galina I. Nesterenko-Stepantsova
Anna S. Katrich
Tatyana I. Penkina
Ekaterina V. Berdnik

Preface

The rapid and impressive results obtained through the application of numerical methods of analysis to electrodynamics created the rather false impression (especially to inexperienced engineers) that all problems were solved long ago or that some of them were not solved because one lacked sufficient time. However, it is sufficient to formulate these problems clearly, to give the task to a programmer to create the corresponding program, and to let a computer "think." It is only left for an engineer to make the corresponding plots and to explain calculation results if the latter do not agree with existing physical notions. However, definite conformity to natural laws starts inevitably to appear while realizing this sequence of steps: the problem is prepared for calculations more thoroughly—the probability to obtain the correct answer is higher, and the obtained information is richer and more interesting—the calculated algorithm is simpler; and as expenditure of calculation time is less, correspondingly, the problems to solve effectively are more complex. The problem formulation is not only compilation of initial equations and a numerical algorithm. It is necessary to foresee a qualitative character of the obtained results and expected order of values, to reveal the peculiarities of the searched solution, complicating its numerical realization and, as a result, choosing the known solution algorithm or working out a new one. If this work has not been done properly, then unexpected anomalies may appear during the numerical calculation process, and the results can be the basis for the most sensational "discoveries." So, the numerical realization is always preceded by a stage of analytical investigation, the efficiency of which depends very much on the degree of mastery of *analytical methods to solve boundary problems* and the ability to apply these methods in a suitable situation, as well as the ability to simplify the initial problem reasonably and also to compare it with problems that have analytical or numerical solutions.

This book contains the original results of the application of analytical methods in the theory of slot-hole coupling of electrodynamic volumes, recently developed by the authors. It is the intent of this book to help its readers acquire this experience.

In Chapter 1, the general equations and general ratios of electrodynamics, boundary conditions for electromagnetic fields, methods of boundary problem solutions, and the method of Green's tensor function creation, which will be used throughout the book, are presented.

Chapter 2 is devoted to the problem set of electromagnetic coupling of electrodynamic volumes via holes, and the initial integral equations have been obtained in it. Here, we also consider the main principles of the asymptotic averaging method, which are applied for the solution of the corresponding boundary problems.

In Chapter 3, the approximate analytical solution of the integral equation concerning the equivalent magnetic current in the narrow slot, coupling two electrodynamic volumes, has been obtained by the averaging method. The formulas and the plots for currents and the coupling coefficients of longitudinal and transverse slots in the common walls of rectangular waveguides are represented.

In Chapter 4, the following electrodynamic structures have been considered: the electrically long longitudinal slot in the common broad wall of rectangular waveguides; two symmetrical transverse slots in the common broad wall of rectangular waveguides; and the transverse slots system in the common broad wall of rectangular waveguides by means of the induced magnetomotive forces method using basis functions of current distribution, obtained by the averaging method.

Chapter 5 is devoted to solving the problem of the resonant iris with the arbitrary oriented slot in the plane of cross section of a rectangular waveguide by the averaging method. Here, we also obtain the formulas for defining the slot resonance wavelength, depending on geometrical parameters of the considered structure.

In Chapter 6, the problem of the stepped junction of two semi-infinite rectangular waveguides with the impedance slotted iris has been solved by the induced magnetomotive forces method. The analytical formulas for the distributed surface impedance of homogeneous and inhomogeneous magnetodielectric coatings of the iris surface have also been obtained.

In Chapter 7, the method of induced magnetomotive forces is used to solve the problem of magnetic current distribution in an antenna–waveguide structure of the "coupling slots in waveguide wall–rectangular resonator–radiating slot" type for the cases of infinite and semi-infinite rectangular waveguides.

For a greater number of the considered electrodynamic structures, the calculated values are compared with the results, obtained by numerical methods (also using commercial programs) and the experimental data.

This book is aimed at academic and industrial researchers involved in microwave techniques, as well as students and postgraduate students studying radiophysics and applied electrodynamics.

The authors consider it their duty to express gratitude to Nikolay A. Khizhnyak and Ludmila P. Yatsuk for their constant attention to and interest in this book; to Nadezhda N. Dyomina and Irina V. Stepantsova for editing the English text; and to Victor M. Dakhov and Svetlana V. Pshenichnaya for help in designing the book.

Mikhail V. Nesterenko
Victor A. Katrich
Yuriy M. Penkin
Sergey L. Berdnik

Contents

Chapter 1
Waves Excitation in Electrodynamic Volumes with Coordinate Boundaries

General equations and general ratios of electrodynamics, boundary conditions for electromagnetic fields, methods of boundary problems solutions and the method of Green's tensor function creation, which will be used throughout the book, are briefly presented in this chapter to facilitate understanding of the manuscript materials by readers. The materials in this chapter, along with commentaries on aspects of electrodynamic theory, will permit a reader to use the book in his of her work without having to track down specialized, largely unavailable, literature.

1.1 Helmholtz's Equations in the Problems of Electrodynamics

Let us consider the oscillations that are harmonic in time, meaning that non-harmonic oscillations can be studied by their expansion in series or Fourier's integral in term of harmonic components. When choosing the dependence from the time t in the form of $e^{i\omega t}$ (ω is the circular frequency of a monochromatic process), the coupling between physical values and their complex amplitudes is determined by the ratios:

$$\vec{E}(t) = \mathrm{Re}\left\{ \vec{E} e^{i\omega t} \right\}, \quad \vec{H}(t) = \mathrm{Re}\left\{ \vec{H} e^{i\omega t} \right\}, \tag{1.1}$$

where Re is the real component of the complex vector in braces; $\vec{E}(t)$ and $\vec{H}(t)$ are the electric field and magnetic field vectors, and $\vec{E}$ and $\vec{H}$ are the complex vector amplitudes, corresponding to them. The complex amplitudes for all physical values included in the electrodynamics equations and oscillating with the $\omega/2\pi$ frequency, are defined in an analogous way.

Let us also introduce the impressed virtual magnetic currents, the densities of which are designated as $\vec{j}^{M}$, alongside with the impressed real electric currents $\vec{j}^{E}$, correspondingly. Then the Maxwell's equations, symmetrical relatively to the electrical and the magnetic values in the absolute Gaussian system of units (CGS),

M.V. Nesterenko et al., *Analytical and Hybrid Methods in the Theory of Slot-Hole Coupling of Electrodynamic Volumes,* DOI: 10.1007/978-0-387-76362-0_1, © Springer Science+Business Media, LLC, 2008

will be written in the following form for the homogeneous and isotropic mediums, the characteristics of which do not change in time:

$$\text{rot } \vec{E} = -ik\mu_1\vec{H} - \frac{4\pi}{c}\vec{j}^M, \quad \text{rot } \vec{H} = ik\varepsilon_1\vec{E} + \frac{4\pi}{c}\vec{j}^E, \tag{1.2}$$

where ε_1 and μ_1 are the complex permittivity and permeability of the medium in which the electric fields are excited, correspondingly; $k = \omega/c$ is the wave number, and $c \approx 2.998 \cdot 10^{10}$ [cm/sec] is the velocity of light in vacuum.

It should be said that the choice of the absolute Gauss system of units as more suitable in theoretical physics is not accidental in this book. Therefore, the Maxwell's equations in the SI units are written in the form of (1.2), non-containing the coefficients 4π and $1/c$. Such a formulation of the main laws of the electromagnetic field leads to the conclusion that four vectors, $\vec{E}$, $\vec{D}$, $\vec{H}$ and $\vec{B}$, defining the field in each spatial point, have different dimensions. Here $\vec{D}$ is the vector of the electrical induction, and $\vec{B}$ is the vector of the magnetic induction. These four vectors are coupled in pairs in vacuum by the ratios: $\vec{D} = \varepsilon_0\vec{E}$ and $\vec{B} = \mu_0\vec{H}$, where the coefficients ε_0 and μ_0 designate the permittivity and permeability, characterizing the electrical characteristics of vacuum, correspondingly. These coefficients of proportionality, lacking real physical meaning loading, are included in all ratios for the fields, and it is difficult to submit with. However, when it is necessary, mutual transition between values representation in the system of CGS and SI is always possible (see Appendix E).

Let us note that the charges are not usually considered. Because of the equation continuity in the theory of monochromatic electromagnetic fields, they are explicitly defined by the currents. In fact, the densities of the impressed charges – both the ρ^E electrical and the ρ^M magnetic ones – are derived from the ratios:

$$\text{div } \vec{j}^E + i\omega\rho^E = 0, \quad \text{div } \vec{j}^M + i\omega\rho^M = 0. \tag{1.3}$$

If one creates the divergences from both parts of the equation (1.2), then, using the identity divrot $\equiv 0$ and the formulas (1.3), we obtain the following equations:

$$\text{div}\left(\varepsilon_1\vec{E}\right) = 4\pi\rho^E, \quad \text{div}\left(\mu_1\vec{H}\right) = 4\pi\rho^M. \tag{1.4}$$

Thus both the charges themselves and the field equations in which they are included do not have independent meaning for the monochromatic field, and that is why they are not taken into consideration. Here we note that the wave equations for the scalar potentials, which are coupled with the vector potentials by the Lorentz's condition (or the gradient invariant condition), are not independently considered further for the same reason.

As it is seen, the equations (1.2) are coupled relatively to the complex vector amplitudes $\vec{E}$ and $\vec{H}$. When solving the majority of the problems of electrodynamics it seems convenient to transit to the equations, in which either the $\vec{E}$ vector or the $\vec{H}$ vector is included:

$$\text{rotrot } \vec{E} = k^2 \varepsilon_1 \mu_1 \vec{E} - \frac{4\pi}{c}\left(ik\mu_1 \vec{j}^E + \text{rot } \vec{j}^M\right),$$
$$\text{rotrot } \vec{H} = k^2 \varepsilon_1 \mu_1 \vec{H} - \frac{4\pi}{c}\left(ik\varepsilon_1 \vec{j}^M - \text{rot } \vec{j}^E\right). \tag{1.5}$$

Having made some standard operations in (1.5) one can transit to the Helmholtz's vector inhomogeneous equations relative to only $\vec{E}$ or $\vec{H}$:

$$\Delta \vec{E} + k^2 \varepsilon_1 \mu_1 \vec{E} = \frac{4\pi}{c}\left(ik\mu_1 \vec{j}^E - \frac{1}{ik\varepsilon_1}\text{graddiv } \vec{j}^E + \text{rot } \vec{j}^M\right),$$
$$\Delta \vec{H} + k^2 \varepsilon_1 \mu_1 \vec{H} = \frac{4\pi}{c}\left(ik\varepsilon_1 \vec{j}^M - \frac{1}{ik\mu_1}\text{graddiv } \vec{j}^M - \text{rot} \vec{j}^E\right), \tag{1.6}$$

where $\Delta \equiv \text{grad div} - \text{rot rot}$ is the Laplace operator.

The obtained equations (1.6) are inconvenient in many cases because of complex representation of their right parts. So, it turned out to be useful to introduce electrodynamic vector potentials of the $\vec{A}^E$ electrical and the $\vec{A}^M$ magnetic types or the $\vec{\Pi}^E$ and $\vec{\Pi}^M$ Hertz's polarization vector potentials (the Hertz's vectors), coupled between each other by the ratios, under the conditions of the accepted suppositions:

$$\vec{A}^E = ik\varepsilon_1\mu_1 \vec{\Pi}^E, \quad \vec{A}^M = ik\varepsilon_1\mu_1 \vec{\Pi}^M. \tag{1.7}$$

The formulas (1.7) underline equivalence of the vector potentials of both types in the problems for the monochromatic fields. As a matter of fact, the choice of one of them over the other is a habit or tradition. Further, we will use the Hertz's vector potentials, with the help of the $\vec{E}$ and $\vec{H}$ vectors, which are expressed as follows in this book:

$$\vec{E} = \text{grad div } \vec{\Pi}^E + k^2 \varepsilon_1 \mu_1 \vec{\Pi}^E - ik\mu_1\text{rot } \vec{\Pi}^M,$$
$$\vec{H} = \text{grad div } \vec{\Pi}^M + k^2 \varepsilon_1 \mu_1 \vec{\Pi}^M + ik\mu_1\text{rot } \vec{\Pi}^E. \tag{1.8}$$

We obtain the following Helmholtz's inhomogeneous equations for the Hertz's vector potentials after the substitution of (1.8) into the equation (1.6):

$$\Delta \vec{\Pi}^E + k^2 \varepsilon_1 \mu_1 \vec{\Pi}^E = -\frac{4\pi}{i\omega\varepsilon_1}\vec{j}^E,$$
$$\Delta \vec{\Pi}^M + k^2 \varepsilon_1 \mu_1 \vec{\Pi}^M = -\frac{4\pi}{i\omega\mu_1}\vec{j}^M. \tag{1.9}$$

As it is seen, simply the excitation currents densities figure in the right parts of the equations (1.9), and that is why the use of these equations is preferable to the use of the equations for the field components (1.6) to solve electrodynamic problems.

To represent the formulas for compactness we will use the following symbols of their designation: $\vec{F}^{E(M)}$ will designate either $\vec{F}^E$ or $\vec{F}^M$; $\varepsilon_1\,(\mu_1)$ will designate either ε_1, or μ_1 further in this book. Then, for example, the equations (1.9) will be written in the form:

$$\Delta\,\vec{\Pi}^{E(M)} + k^2\varepsilon_1\mu_1\vec{\Pi}^{E(M)} = -\frac{4\pi}{i\omega\varepsilon_1\,(\mu_1)}\,\vec{j}^{E(M)}. \tag{1.10}$$

Let us note that the equations written above did not try "to impose" any conditions on the coordinates system choice (except the condition of their orthogonality, providing the possibility of using differential operators apparatus). However, it is useful to underline one special classical application for the Hertz's vectors in the spherical coordinates system (r, θ, φ) here. It concerns their coupling with the Debye's potentials [1.1]. It is expressed by the following ratios at the impressed current's absence:

$$\vec{\Pi}^E = \vec{r}u + \mathrm{grad}\psi^E, \qquad \vec{\Pi}^M = \vec{r}v + \mathrm{grad}\psi^M, \tag{1.11}$$

where $\vec{r} = \vec{r}^0 r$ ($\vec{r}^0$ is the unit vector); u and v are the Dabye's potentials of the electrical and magnetic types, satisfying the Helmholtz's homogeneous equations of the type (1.10), correspondingly; $\psi^{E(M)}$ are the scalar functions, satisfying the equations:

$$\tilde{\Delta}\,\psi^E + k^2\varepsilon_1\mu_1\psi^E + 2u = const, \qquad \tilde{\Delta}\,\psi^M + k^2\varepsilon_1\mu_1\psi^M + 2v = const, \tag{1.12}$$

where $\tilde{\Delta} \equiv \mathrm{div}\,\mathrm{grad}$ is the scalar functions Laplacian.

The solutions of the equations (1.2) are not usually investigated independently because the electromagnetic field value, which is expressed in this case as:

$$\vec{E} = \mathrm{rotrot}\,(\vec{r}u) - i\omega\mu_1\mathrm{rot}\,(\vec{r}v)\ , \qquad \vec{H} = \mathrm{rotrot}\,(\vec{r}v) + i\omega\varepsilon_1\mathrm{rot}\,(\vec{r}u)\ , \tag{1.13}$$

does not depend on the $\psi^{E(M)}$ function.

1.2 Boundary Conditions for Electromagnetic Fields

The parameters of the real mediums ε_1 and μ_1, figuring in the equations as constants, considered in a Sect. 1.1, can vary in space arbitrarily. This change can occur both continually and in discrete steps. For example, the parameters ε_1 and μ_1 can have discontinuities on a surface —such a surface is the interface of two mediums. Generally speaking, the electromagnetic fields vectors also have discontinuities on a surface S, which is why the equations (1.2) must be added by the boundary conditions (the adjunction conditions for the electromagnetic field vectors). As it is known from classical literature these conditions have the following form on smooth interfaces:

$$\mu_1 \left(\vec{H}_1, \vec{n} \right) - \mu_2 \left(\vec{H}_2, \vec{n} \right) = 4\pi \rho_S^M, \tag{1.14}$$

$$\varepsilon_1 \left(\vec{E}_1, \vec{n} \right) - \varepsilon_2 \left(\vec{E}_2, \vec{n} \right) = 4\pi \rho_S^E, \tag{1.15}$$

$$\left[\vec{n}, \vec{E}_1 \right] - \left[\vec{n}, \vec{E}_2 \right] = \frac{4\pi}{c} \vec{j}_S^M, \tag{1.16}$$

$$\left[\vec{n}, \vec{H}_1 \right] - \left[\vec{n}, \vec{H}_2 \right] = -\frac{4\pi}{c} \vec{j}_S^E, \tag{1.17}$$

where indices 1 and 2 correspond to the first and the second mediums, $\rho_S^{E(M)}$ are the surface charges densities, $\vec{j}_S^{E(M)}$ are the surface currents densities, $\vec{n}$ is the unit normal vector to the interface, directed into the second medium. Let us note that these conditions are directly derived from Maxwell's equations (1.2).

When the set impressed electrical currents are absent on the S surface of discontinuity, if S is not perfectly conducting, the condition (1.17) has the form:

$$\left[\vec{n}, \vec{H}_1 \right] = \left[\vec{n}, \vec{H}_2 \right] \tag{1.18}$$

and designates continuity of the magnetic field tangential component on the given surface. Analogously, if $\vec{j}_S^M = 0$, then the condition (1.16) designates the tangential component continuity of the magnetic field on the mediums interface. Equations (1.14) and (1.15) describe the values change of normal components of the electromagnetic field on the mediums interface.

If the second medium has infinite conductivity (that is, the interface is perfectly conducting), then the fields $\vec{E}_2$ and $\vec{H}_2$ are identically equal to zero, and the boundary condition must be performed on the perfectly conducting body surface:

$$\left[\vec{n}, \vec{E}_1 \right] \Big|_{\text{on } S} = 0. \tag{1.19}$$

The body's conductivity is large but finite in many real cases. The precise boundary conditions (1.16) and (1.17) must be performed on the surfaces of such bodies, which means the Maxwell's equations must be solved both outside and inside the body with the adjunction conditions on the interface to define the electromagnetic field outside a perfectly conducting body in this case. Such a problem is more complex than the problem of the electromagnetic field definition in one of the mediums with the set boundary condition on the surface. So, it is preferable to change the adjunction conditions on the body surface on the boundary condition, coupling the field vectors values on the surface only in one of the mediums. Such boundary conditions are the Leontovich–Shchukin approximate boundary conditions or the impedance boundary conditions [1.2]:

$$\left[\vec{n}, \vec{E} \right] = -\bar{Z}_S \left[\vec{n}, \left[\vec{n}, \vec{H} \right] \right] \Big|_{\text{on } S}, \tag{1.20}$$

where $\vec{n}$ is the normal to the body surface, directed inside the body, and $\bar{Z}_S = Z_S/Z_0$ (normalized to the free space impedance $Z_0 = 120\pi$ Ω) is the distributed complex impedance on the S surface. The $\bar{Z}_S$ impedance can be defined in a matrix form (for an anisotropic surface) in a general case.

Let us underscore that the impedance condition (1.20) is approximate in the sense that the problem solved with its use is the first term of asymptotic decomposition of a precise solution of the adjunction small parameter problem $|Z_S/Z_0| << 1$.

For the community of the problem account it should be noted that the condition (1.20) occurs for a medium's plain interface or such an interface, the radiuses of curvature of which are much larger than the falling wavelength. More general conditions, taking into consideration the interface surface curvature, are the conditions of the form:

$$E_{\tau 1} = \bar{Z}_S \left(1 + \frac{\chi_1 - \chi_2}{2ik_2}\right) H_{\tau 2}\bigg|_{\text{on } S}; \quad E_{\tau 2} = -\bar{Z}_S \left(1 + \frac{\chi_2 - \chi_1}{2ik_2}\right) H_{\tau 1}\bigg|_{\text{on } S},$$
$$(1.21)$$

where χ_1 and χ_2 are the main Gaussian curvatures of the S surface, E_τ and H_τ are the tangential components of the electromagnetic fields on this surface, located in the corresponding planes, and $k_2 = \frac{\omega}{c}\sqrt{\varepsilon_2\mu_2}$ is the wave number in the adjacent layer, characterized by the $\bar{Z}_S$ impedance.

We note here, that it turns out to be right to use "traditional" forms of the impedance boundary conditions (1.20) for the S sphere surface, which are umbilical surfaces ($\chi_1 = \chi_2$).

Because the electromagnetic field must be defined in whole space, it is necessary to demand the fulfillment of the field finiteness (boundedness) in any point of the spatial considered region and also the conditions on the infiniteness for its uniqueness, except the conditions on the medium's interfaces. As it is known, the conditions on infiniteness must satisfy known physical requirements: they must extract the waves going into infiniteness and accept the possibility of waves formation coming from infiniteness. At this the fields on infiniteness decrease faster, than $1/R$, where R is the distance from the fixed point.

1.3 Application Peculiarities of the Theorem of Uniqueness and the Principle of Duality for the Regions with Impedance Boundaries

Because the electrodynamic problems with the use of impedance boundary conditions will be investigated further in the book, it is useful to discuss some "gentle" aspects of such boundary conditions' use.

1.3.1 Peculiarities of Application of the Theorem of Uniqueness

The problems, which we have to solve in the theory of monochromatic electromagnetic fields, are divided into two categories: inner problems and outer problems. The field in the bounded spatial portion, surrounded by the S surface, is considered in the inner problems. The $\vec{j}^{E(M)}$ impressed currents are set inside the S surface, and the boundary conditions for the tangential components of the E_τ electric and the H_τ magnetic fields—on the surface itself. The fields in the unbounded space outside some S surface, on which the boundary conditions for the E_τ and H_τ components are also set, and the impressed fields set in the space surrounding this surface, are considered in the outer problems.

From a mathematical point of view the requirement to perform one or another boundary conditions on S from Section 1.2 must define its unique solution in both problems for the Helmholtz's equations (1.6) or (1.10). This statement is proved in the theory of electrodynamics within the frame of the theory of uniqueness.

The problems of uniqueness of solution of inner and outer boundary problems of the Maxwell's equations system have been investigated for the monochromatic electromagnetic fields (let us choose the temporal dependence in the form of $e^{-i\omega t}$ for the convenience to compare with literary sources here) in the conducting mediums, which are characterized by the complex values of the $\varepsilon_1 = \mathrm{Re}\,\varepsilon_1 + i\,\mathrm{Im}\,\varepsilon_1$ permittivity and the $\mu_1 = \mathrm{Re}\,\mu_1 + i\,\mathrm{Im}\,\mu_1$ permeability ($\mathrm{Im}\,\varepsilon_1 > 0$, $\mathrm{Im}\,\mu_1 > 0$), varying from one point to another one in the known literature in general. These problems have been sufficiently studied in the cases of the interfaces, on which the boundary conditions for the electric and the magnetic fields are set separately from each other [1.2]. The proving aspects of the theorem of uniqueness have been elucidated contradictorily in the literature, which demands their additional consideration in [1.3] for the boundary problems with the impedance surfaces, where the boundary conditions (1.20) couple the fields tangential components by the functional dependence.

The lemma used, that the Maxwell's homogeneous equations do not have the distinctive from the null solution, is a dominant moment at the proof of the theory of the uniqueness in any case. It is applied for the difference electromagnetic fields $\left(\vec{E}, \vec{H}\right)$ and is proved either on the basis of the Poynting's complex theorem [1.2] or on the basis of the Lorentz's lemma application [1.3].

If the tangential component of the electrical field (defined on the S_1 part of the S surface) and the magnetic field tangential component (defined on the S_2 part of the surface $S = S_1 + S_2$) are set on the S boundary surface in an evident form, the difference field $\left(\vec{E}, \vec{H}\right)$ must satisfy the Maxwell's homogeneous equations with the homogeneous boundary conditions on the S surface: $E_\tau = 0$ on S_1 and $H_\tau = 0$ on S_2.

Likewise, if the Leontovich–Shchukin impedance boundary conditions (1.20) are set on the S surface, it is not difficult to assure that the difference field $\left(\vec{E}, \vec{H}\right)$ will also satisfy the Maxwell's homogeneous linear equations but with the impedance boundary conditions (1.20) on the S surface. This difference dictates the necessity of a separate proof of the theorem of uniqueness for the boundary problems with impedance surfaces in the obtained boundary conditions for the difference fields.

One should say that different approaches are used to define the surface impedance in the literature. So, this is the linear operator, coupling the tangential components $\vec{E}_\tau$ and $\left[\vec{H}_\tau, \vec{n}\right]$ in some monographs, and it is defined from the coupling $\vec{E}_\tau$ and

$\left[\vec{n}, \vec{H_\tau} \right]$ in other works. That is why it is always necessary to observe the content of the used surface impedance definition clearly.

Let us designate the anisotropic surface impedance in the form of the matrix for a general case:

$$\hat{Z}_S = \left\| \begin{matrix} Z_{11} & Z_{12} \\ Z_{21} & Z_{22} \end{matrix} \right\|, \tag{1.22}$$

where $Z_{jq} = R_{jq} + i X_{jq}$, j and q are the indices, taking the values 1,2. Let us note that at this the performance of a series of inequations is required:

$$R_{11} \geq 0; \quad R_{22} \geq 0; \quad 4R_{11} R_{22} \geq \left| Z_{12} + Z_{21}^* \right|^2, \tag{1.23}$$

where Z_{21}^* is the complex conjugated values of the Z_{21} component. These inequations provide absence of the additional energetic sources, more accurately, the energy flux via the surface inside the region in question on the impedance surface. Let us underscore that physically correct use of the impedance boundary conditions in the form of (1.20), when $\vec{n}$ is the normal vector to the S surface, is directed into "the impedance body", and is possible only at the fulfillment of the requirements (1.23) in the case of the $\hat{Z}_S$ anisotropic surface impedance or at the performance of the condition of Re $Z_S \geq 0$ in the case of the Z_S isotropic surface impedance.

Supposing that not only electrical but magnetic losses with the use of the impedance boundary condition (1.20) on the S set surface differ from null in each point of the D_e spatial region in question, the proof of the theorem of uniqueness has been done by the authors with the help of the Lorentz's lemma application to the difference electromagnetic field $\left\{ \vec{E}_d, \vec{H}_d \right\} = \left\{ \vec{E}, \vec{H} \right\} - \left\{ \vec{E}^*, \vec{H}^* \right\}$, where * is the complex conjugated token. We represent the final expression for the case of the Z_S isotropic surface impedance here:

$$\oint_S \text{Re } Z_S \left| \left[\vec{n}, \vec{H}_d \right] \right|^2 ds + \omega \int_{D_e} \left(\text{Im } \varepsilon_1 \left| \vec{E}_d \right|^2 + \text{Im } \mu_1 \left| \vec{H}_d \right|^2 \right) dv = 0, \tag{1.24}$$

where dv is the volume element. As can be seen, it contains the integral items, stipulated by the medium electrical and magnetic losses consideration and the surface integral, the value of which is proportional to Re Z_S. It is obvious that to perform the condition (1.24), $\vec{E}_d = 0$ and $\vec{H}_d = 0$ must be everywhere in the D_e region at Im $\varepsilon_1 > 0$ and also when Im $\mu_1 > 0$. Let us note that the expression (1.24) also turns out to be just in the case of the time dependence choice of monochromatic wave processes in the $e^{i\omega t}$ form. At this, both token changes simultaneously at ω and of Im ε_1 and Im μ_1 in (1.24).

Thus we can state that solutions of the inner and outer boundary problems of the Maxwell's equations system in the regions, limited by impedance surfaces, are unique at the Re$Z_S \geq 0$ condition fulfillment in the case of the isotropic surface impedance and the requirements performance (1.23) in the case of the anisotropic impedance surface in the conducting mediums, for which the $\text{Im}\varepsilon_1 > 0$ and $\text{Im}\mu_1 > 0$.

1.3.2 Peculiarities of Application of the Principle of Duality

The principle of duality (or the theorem of duality) usually used in Ya. Feld formulation [1.2], sets the coupling between two diffraction problems: 1) the diffraction in free space on a plain infinitely thin and perfectly conducting plate; 2) the diffraction in the free space on a plain infinitely thin and perfectly conducting screen, having the hole, which reproduces the plate in the first problem by forms and sizes. We note that the theorem of duality is the analogue of Babinet's theorem, known in physical optics, which couples the diffraction phenomena on the mutual adding screens in the frame of the Huygens's scalar principle. The analysis of the possibility to apply the principle of duality for the impedance surfaces regions was absent in the literature before the monograph [1.3].

The principle of duality is grounded in the characteristics of permutation symmetry or permutation invariance of Maxwell's equations with respect to the $\vec{j}^E$ electric and the $\vec{j}^M$ magnetic impressed currents. It follows from these characteristics that the following two-sided changes are possible:

$$\vec{j}^E \Leftrightarrow -\vec{j}^M; \quad \vec{H}_1 \Leftrightarrow \vec{E}_2; \quad \vec{E}_1 \Leftrightarrow \vec{H}_2; \quad \varepsilon_1 \Leftrightarrow -\mu_1. \tag{1.25}$$

One can ground the choice of one or another pair of coupled electrodynamics problems of concrete geometry, differing by the conditions of excitation $\left(\vec{j}^E \Leftrightarrow -\vec{j}^M \right)$, which will define the content of the principle of reciprocity for each of possible variants on the basis of analysis. One should note that the necessity to use the Maxwell's equations, which are non-symmetrical relative to the impressed currents, occurs when considering the problems only with one definite type of the currents of excitation. Direct use of permutation invariance characteristics is difficult in these cases, and separate proofs of the principle of duality are needed as, for example, in the monograph [1.2].

The two-sided permutations (1.25) are just for the electrodynamics problems, investigated in infinite space at the fulfillment of the field boundedness conditions on infinity. The use of (1.25) in the boundary problems with the impedance surfaces will have one essential peculiarity. It concludes that mutual permutations must also transform the boundary conditions in a corresponding way.

Because the impedance boundary conditions (1.20) are non-symmetrical relative only to the change of the fields $\vec{E}_1 \Leftrightarrow \vec{H}_2$ and $\vec{H}_1 \Leftrightarrow \vec{E}_2$, it is also necessary to change the surface impedance values. It is not difficult to assure that permutations (1.25) must be added by one more in the case of the Z_S isotropic impedance:

$$Z_S \Leftrightarrow -1/Z_S. \tag{1.26}$$

Rightness of the requirements (1.26) is simpler to analyze on the example of the problem about normal incidence of the plain electromagnetic wave on the medium's interface plain: empty half-space and the half-space with homogeneous medium, characterized by the complex permittivity ε_1 and permeability μ_1. The notion of the wave impedance of medium in the form of $\bar{Z}_S = \sqrt{\mu_1/\varepsilon_1}$ is traditionally defined, namely, from the solution of this problem. The requirement (1.26) becomes evident at the field's change due to (1.25) and $\varepsilon_1 \Leftrightarrow -\mu_1$.

When analyzing the permutation (1.26) one can make the following conclusion: the principle of duality can be realized in practice only at the following (1.26) change of the impedance structures in coupled problems in any electrodynamic volumes containing the impedance boundaries. Because the Re $Z_S > 0$ condition is the condition of physical realization of passive impedance structures, this change is possible only for the surfaces characterized by purely imaginary impedances.

This conclusion also stays fair in the case of the surface impedance representation in the matrix form (1.22). The condition (1.26) can be written in the following form, considering that matrix algebra coincides with linear operators algebra:

$$\hat{Z}_S \Leftrightarrow -\hat{Z}_S \Big/ \det\left[\hat{Z}_S\right], \tag{1.27}$$

where $\det\left[\hat{Z}_S\right]$ is the $\hat{Z}_S$ impedance matrix determinant. At this one should note that the permutation (1.27) is possible, only when $Z_{12} = Z_{21}$, that is, the $\left[\hat{Z}_S\right]$ matrix is symmetrical.

1.4 Application of the Method of Moments in the Coupling Slot-Holes Theory

The Helmholtz's vector equations from Sect. 1.1 create a unique boundary problem, mathematically formulated in the form of the integral–differential equations or their systems, jointly with the concrete boundary conditions of Sect. 1.2. We have to solve these equations relative to the $\vec{j}^{E(M)}$ unknown currents by different methods in the theory of linear antenna radiators. And though this book is devoted mainly to the numerical–analytical approaches to the solution of such equations, we often have to combine them with another numerical methods application. Therefore, we decided it would be expedient to consider the application of the numerical method of moments briefly in the coupling slot-holes theory in one key here. We note that the achievements of the theory of perfectly conducting thin vibrators are successfully "used" due to the principle of reversibility of slot element in this problem.

Let us consider the equation $\hat{L}^{E(M)}(\vec{r}, \vec{r}')\vec{j}^M(\vec{r}') = \vec{u}^{E(M)}(\vec{r})$, where $\hat{L}^{E(M)}$ $(\vec{r}, \vec{r}')$ is the tensor integral–differential operator in a general case, and $\vec{j}^M(\vec{r}')$ is the magnetic current, equivalent to the $\vec{E}$ electrical field in the slot element aperture as a model problem for coupling slot-holes. The method of moments is more universal in different modifications among approximate numerical methods of the solution of the equations of such a kind. Its presentation is given, for example, in the monograph [1.4] in application to the problems of electrodynamics. The essence of the method of moments to solve the integral–differential equations is sufficiently simple: it consists of reduction of the equation in question to the system of linear algebraic equations (SLAE). For this the unknown function is represented in the form of linear combinations of the known functions with unknown complex amplitudes, and it is substituted into the original equation. That is, the equivalent magnetic currents solution is searched in the form of: $\vec{j}^M(\vec{r}') = \sum_m C_m \vec{\varphi}_m(\vec{r}')$ where C_m are the unknown complex amplitudes, and $\vec{\varphi}_m(\vec{r}')$ are the vector basic functions, which must create a part of a system of linearly independent functions. If the system of the $\{\vec{\varphi}_m(\vec{r}')\}$ functions is complete in the functional space we are interested in, then we can reach the precise solution of the set problem at $m \rightarrow \infty$. To obtain the approximate solution, M = max (m) is chosen to be finite on the basis of preliminary grounds.

The right part of the equation is also represented in the form of the sum: $\vec{u}^{E(M)}(\vec{r}) = \sum_\mu u_\mu \vec{\psi}_\mu(\vec{r})$, where the $\vec{\psi}_\mu(\vec{r})$ vector functions are called weighty or testy, and L = max(μ). They must also be linearly independent. Further problem solution can be obtained in several stages: 1) it is necessary to multiply scalarly both parts of the equation on the $\vec{\psi}_\mu(\vec{r})$ weighty functions; 2) one must integrate both parts of the equation along the slot element surface. As a result the inner products are calculated, and by this the equations are reduced to the SLAE matrix form; 3) SLAE solution is obtained for the C_m unknown complex amplitudes.

SLAE is obtained from (2L + 1) of the equations with (2M + 1) by the C_m unknown amplitudes on the 2nd stage of the solution. The SLAE matrix turns out to be square at M = L, and the C_m amplitudes are obtained by the solution of the linear algebraic equations system by any method. In the case, when the vector weighty functions coincide with the basic functions of ($\vec{\psi}_\mu = \vec{\varphi}_m$), the method of moments is called Galerkin's method. One should note that the solution obtained by Galerkin's method has stationary characteristics, and the method itself is equivalent to the Rayleigh–Ritz's variational method [1.4].

The method of moments is known as the generalized method of induced electromotive forces (EMF) in the theory of vibrator radiators. This term is connected with the development of the antennas theory, which is a part of radio engineering. The unknown function in the initial integral equation in electrodynamics and the antennas theory is usually an electrical current, and the kernel of the equation has the sense of a resistance. So the inner products (the matrix coefficients) represent themselves as an electromotive force, induced in a given place on the antenna by currents of its other sections or other radiators. The right part of the original equation usually corresponds to the impressed EMF. The term "the generalized method of the induced EMF" successfully reveals the physical sense of the used formulas. If the equations for the magnetic currents, equivalent to the slot field, are solved, then the matrix coefficients in SLAE have admittances sense, and the method is called the "generalized method of induced magnetomotive forces (MMF)".

The basic functions choice has paramount significance in the method of moments. Infinitely large number sets of basic and weighty functions exist principally. However, it is expedient to choose these functions so that they correspond to physical essence and lead to effective calculating algorithms for every concrete electrodynamics problem. The kinds of basic and weighty functions or the kinds of spatial harmonics of current and voltage for antenna elements can be divided into two groups conditionally: 1) the harmonics, defined on the whole element (the basic functions of a complete region); 2) the harmonics, differing from null only within an element region (the basic functions of sub-regions or the segmental basic functions).

Let us represent the often used harmonics of the 1st group:

- *Harmonics of the Fourier series* $J(u) = \sum_m C_m e^{\frac{2\pi i m u}{T}}$. The apparatus of the Fourier series is a classical apparatus of approximations. Its characteristics have been studied properly. The formulas used to obtain SLAE coefficients are relatively simple, and it has accuracy estimation at the uniform and mean–square approximations. The harmonics of the Fourier series are often used to represent the currents and the voltages in rectilinear antenna elements. When the current on the antenna element is transformed into null on its edges, it is natural to take the expansion in terms of cosine functions, if the beginning of the coordinates coincides with the middle of the element;

- *Power polynomials* $J(u) = \sum_m C_m u^m$. The power polynomials apparatus also refers to classical means of approximation functions. The interpolation formulas and estimations of accuracy are well-known. It is convenient to use the Dirac delta function as weighty functions when using power polynomials as the basic functions, that is, to make point-wise lacing. It allows us to get rid of one from two integral at mutual admittances (impedances) calculations [1.5];

- *Chebyshev polynomials* $J(u) = \sum_m C_m T_m(u)$. They are applied as the basic functions to represent the currents in antenna elements, where a strong edge effect is observed (for example, the aperture and strip line ones). Sharp increase of the Chebyshev polynomials values on the edges of the set integral permits us to describe the edge effect (the element edge diffraction) correctly, and it speeds up the convergence of the precise solution;

- *Bessel functions* $J(u) = \sum_m C_m J_m(u)$. These functions are sufficiently studied, and standard programs are available for their calculations. They are usually used as basic functions in elements with axial symmetry;

- *King's trigonometrical harmonics*. As for the authors' statement, the current representation by the sum of the trigonometric functional trinomial combination, gives rather close to precise distribution of the current [1.6].

Let us consider some harmonics of the 2nd group, which can be represented in a general form as:

$$\varphi_m(u) = \begin{cases} f_m(u) \text{ at } u \in \Delta u_m \\ 0 \text{ in the rest of the region} \end{cases}.$$

Such kinds of harmonics of sub-regions are used much more often:

- *Dirac delta function* $J(u) = \sum_m C_m \delta(u - u_m)$. The choice of this basis means the boundary conditions satisfaction in separate local points of the antenna element, which is called point-wise lacing or the collocation method. Due to the delta-functional characteristics, all integrals disappear, and calculation of mutual admittances (impedances) and the right parts of SLAE becomes the simplest. Many examples of successful application of such bases are known. However, one has to take into account rather many lacing points and must often solve SLAE of higher orders. It is necessary to evaluate the compromise between simplicity of matrix calculating algorithms of the system and complexity of SLAE solution with large numbers of equations at concrete problems analysis;
- *Step functions* $f_m(u) = I_m$. This basis is close to the previous one due to its characteristics. Its uniqueness and simplicity are compensated by less rapid convergence to a precise solution and difficulty of SLAE solution of higher order at integration;
- *Piece-linear functions* $f_m(u)\Delta u_m = I_m(u_{m+1} - u) + I_{m+1}(u - u_m)$. This basis speeds up convergence to a precise solution with respect to the previous ones at rather simple calculation of the integrals;
- *Piece-sinusoidal functions* (the Richmond's functions [1.4]) $f_m(u) \sin k\Delta u_m = I_m \sin k(u_{m+1} - u) + I_{m+1} \sin k(u - u_m)$. The advantages of this basis are in relatively small volume of calculations, because closed expressions can be obtained for many integrals, and rapid convergence to a precise solution;
- *Parabolic functions* $f_m(u) = A_m + B_m(u - u_m) + D_m(u - u_m)^2$. The subregions can cross and the approximation character, in its essence, is close to the approximation by splines here.

1.5 Green's Tensor Functions of the Helmholtz's Vector Equation for the Hertz's Potentials

1.5.1 Green's Tensor Function Characteristics

As was indicated earlier, the Helmholtz's vector equations from Sect.1.1 make a unique boundary problem together with the boundary conditions from Sect.1.2. Two methods for analytically solving such problems are known in the literature: the method of eigenfunctions and the Green's function method.

The method of the eigenfunctions, for example [1.2], is based on the differential equations solution with the help of separation of the variables. We remind that solutions of an ordinary homogeneous differential equation containing the constant of separation, are called eigenfunctions, which satisfy the corresponding boundary conditions on the edges of the region variation of independent variables. The values of the constant of separation, assumed under the prescribed conditions, are called eigenvalues. This method is not used in the present book, which is why its application is not considered in detail here.

The second method is more physically evident, as the Green's functions represent themselves the field (in a general sense) in the observation point, created by a unique point source. To obtain the field created by the combination of the sources distributed in space with a definite density, one should obtain the corresponding volume integral in the whole region where the sources are set. Therefore, it is expedient to use the Green's function method, namely, at the solution of the problems of spatial regions excitation by the impressed currents.

Let us consider the main characteristics of the Green's tensor functions [1.7] of the Helmholtz's vector equation, which will be used further:

a) the Green's tensor function $\hat{G}\left(q_1, q_2, q_3, q_1', q_2', q_3'\right) = \hat{G}\left(\vec{q}, \vec{q}'\right)$ satisfies the Helmholtz's inhomogeneous equation in the orthogonal curvilinear coordinates system (q_1, q_2, q_3):

$$\Delta\hat{G}\left(\vec{q}, \vec{q}'\right) + k^2\,\hat{G}\left(\vec{q}, \vec{q}'\right) = -4\pi\,\hat{I}\,\frac{\delta\left(q_1 - q_1'\right)\delta\left(q_2 - q_2'\right)\delta\left(q_3 - q_3'\right)}{h_1 h_2 h_3},\quad (1.28)$$

where $\hat{I}$ is the unit tensor (dyadic); $\left(q_1', q_2', q_3'\right)$ are the source point coordinates; $\delta\left(q - q'\right)$ is the Dirac delta function; h_n are the Lamé coefficients. The Laplacian is applied to all components of the tensor in (1.28);

b) The Green's function is the symmetrical tensor of the second rank—affinor, defined by nine components. The performance of the equality provides tensor symmetry:

$$\vec{F}\left(\vec{q}\right)\hat{G}\left(\vec{q}, \vec{q}'\right) = \hat{G}\left(\vec{q}, \vec{q}'\right)\vec{F}\left(\vec{q}\right).\quad (1.29)$$

Availability of nine components is stipulated physically by that the fact that each of three components of the vector source can create three field components in a general case;

c) All components of Green's tensor are invariant relative to the interchange of the coordinates of the $\left(\vec{q}'\right)$ source points and the $(\vec{q})$ observation ones;

d) All components of Green's tensor are characterized by integrated peculiarity of the following type at coincidence of the source points and the $\vec{q} \to \vec{q}'$ observation ones:

$$1 \Big/ \sqrt{(q_1 - q_1')^2 + (q_2 - q_2')^2 + (q_3 - q_3')^2} \, ;$$

e) One can obtain the solution of the Helmholtz's vector inhomogeneous equation in an integral form with the help of Green's tensor.

This solution is represented in the following form for $\vec{\Pi}^{E(M)}$ Hertz's vector potentials due to [1.7]:

$$
\begin{aligned}
\vec{\Pi}^{E(M)}(\vec{q}) = {}& \frac{1}{i\omega\varepsilon_1(\mu_1)} \int\limits_V \vec{j}^{E(M)}(\vec{q}') \, \hat{G}^{E\,(M)}(\vec{q}, \vec{q}') \, dv' \\
& + \oint\limits_S \Big\{ \mathrm{div}\vec{\Pi}^{E(M)}(\vec{q}') \, \hat{G}^{E(M)}(\vec{q}, \vec{q}') \, \vec{n} - \mathrm{div}\hat{G}^{E(M)}(\vec{q}, \vec{q}') \, \vec{\Pi}^{E(M)}(\vec{q}') \, \vec{n} \\
& + \Big[\vec{n}, \, \hat{G}^{E(M)}(\vec{q}, \vec{q}') \Big] \mathrm{rot}\, \vec{\Pi}^{E(M)}(\vec{q}') - \Big[\vec{n}, \, \vec{\Pi}^{E(M)}(\vec{q}') \Big] \mathrm{rot}\hat{G}^{E(M)}(\vec{q}, \vec{q}') \Big\} ds',
\end{aligned}
$$

$$(1.30)$$

where $\vec{n}$ is the ort of the outer normal to the S surface. The volume integral is taken over the whole volume, bounded by S (dv is the volume element) and the surface integral—over the whole S surface (ds$'$ is the area element in the $\vec{q}'$ primed coordinates); the differential operations are also performed due to the $\vec{q}'$ primed coordinates in (1.30). We note that the expression in brace brackets is a vector.

Thus, solution of Helmholtz's inhomogeneous equation is a sum of the volume and surface integrals. The integration elements of the surface integrals contain the boundary values of the unknown function and its derivatives, which must be known beforehand. Of course, it limits the circle of the solved problems. If the boundary values of the unknown function are not known beforehand, the equalities (1.30) transform into the integral equations, the solution of which is no less complex than the solution of the original differential equations.

We can neglect the surface integrals in (1.30), creating Green's function by special means. If the $\hat{G}(\vec{q}, \vec{q}')$ Green's function components satisfy the same boundary conditions on the S surface as the corresponding components of the $\vec{\Pi}^{E(M)}(\vec{q})$ vector potentials, the surface integrals disappear due to the fact that vector components of the integration element transform into zero. Therefore, the solution for $\vec{\Pi}^{E(M)}(\vec{q})$ will be represented only in the form of the volume integral from (1.30) in this case.

Having substituted this solution into the equation (1.8), we obtain the equalities, from which the electromagnetic field intensities will be defined in the following form:

$$\vec{E}\,(\vec{q}) = \frac{1}{4\pi i \omega \varepsilon_1} \left[\text{graddiv} + k_1^2 \right] \int_V \vec{j}^E\,(\vec{q}\,') \, \hat{G}^E\,(\vec{q}, \vec{q}\,') \, dv'$$

$$- \frac{1}{4\pi} \text{rot} \int_V \vec{j}^M\,(\vec{q}\,') \, \hat{G}^M\,(\vec{q}, \vec{q}\,') \, dv',$$

$$\vec{H}\,(\vec{q}) = \frac{1}{4\pi i \omega \mu_1} \left[\text{graddiv} + k_1^2 \right] \int_V \vec{j}^M\,(\vec{q}\,') \, \hat{G}^M\,(\vec{q}, \vec{q}\,') \, dv' \qquad (1.31)$$

$$+ \frac{1}{4\pi} \text{rot} \int_V \vec{j}^E\,(\vec{q}\,') \, \hat{G}^E\,(\vec{q}, \vec{q}\,') \, dv',$$

where $k_1 = k\sqrt{\varepsilon_1 \mu_1}$.

Using the representation of the electromagnetic field via the Green's field dyadic function [1.8]:

$$\begin{bmatrix} \vec{E}\,(\vec{q}) \\ \vec{H}\,(\vec{q}) \end{bmatrix} = \int_V \begin{bmatrix} \hat{g}^E\,(\vec{q}, \vec{q}\,') & \hat{g}^{EM}\,(\vec{q}, \vec{q}\,') \\ \hat{g}^{ME}\,(\vec{q}, \vec{q}\,') & \hat{g}^M\,(\vec{q}, \vec{q}\,') \end{bmatrix} \cdot \begin{bmatrix} \vec{j}^E\,(\vec{q}\,') \\ \vec{j}^M\,(\vec{q}\,') \end{bmatrix} dv', \qquad (1.32)$$

where $\hat{g}^E\,(\vec{q}, \vec{q}\,')$ is the electrical, $\hat{g}^M\,(\vec{q}, \vec{q}\,')$ is the magnetic, $\hat{g}^{EM}\,(\vec{q}, \vec{q}\,')$ is the electromagnetic, $\hat{g}^{ME}\,(\vec{q}, \vec{q}\,')$ is the magnetoelectric functions, one can transit to the representations for these functions, defined through the Green's tensors of the Hertz's vector potentials, by comparison of the expressions (1.31) and (1.32):

$$\hat{g}^E\,(\vec{q}, \vec{q}\,') = \frac{1}{4\pi i \omega \varepsilon_1} \left[\text{graddiv} + k_1^2 \right] \hat{G}^E\,(\vec{q}, \vec{q}\,');$$

$$\hat{g}^{EM}\,(\vec{q}, \vec{q}\,') = -\frac{1}{4\pi} \text{rot} \hat{G}^M\,(\vec{q}, \vec{q}\,');$$

$$\hat{g}^{ME}\,(\vec{q}, \vec{q}\,') = \frac{1}{4\pi} \text{rot}\, \hat{G}^E\,(\vec{q}, \vec{q}\,'); \qquad (1.33)$$

$$\hat{g}^M\,(\vec{q}, \vec{q}\,') = \frac{1}{4\pi i \omega \mu_1} \left[\text{graddiv} + k_1^2 \right] \hat{G}^M\,(\vec{q}, \vec{q}\,').$$

As it was indicated above, the Green's functions for the vector potentials $\hat{G}^{E(M)}\,(\vec{q}, \vec{q}\,')$ have integral singularity at $\vec{q} \rightarrow \vec{q}\,'$ in the source region. It turns out to be non-integrable in the case of the Green's field functions. To analyze the latter one has to use the apparatus of the theory of the generalized functions with the purpose of its regularization, which complicates the problem of the Green's tensor creation considerably.

In connection with this, the problems solution of excitation of different electrodynamic volumes by the impressed fields will be obtained with the help of the Green's tensor function of the Helmholtz's equations for the Hertz's vector potentials $\vec{\Pi}^{E(M)}\,(\vec{q})$ later in the book. The intensities of the $\vec{E}\,(\vec{q})$ and $\vec{H}\,(\vec{q})$ electromagnetic field will be obtained from the ratios (1.31) after definition of $\vec{\Pi}^{E(M)}\,(\vec{q})$.

1.5.2 Method of the Green's Tensor Creation in the Orthogonal Curvilinear Coordinates Systems

The method of creation of the Green's tensor function of the Helmholtz's vector equation, developed in the monograph [1.7], is grounded on the possibility of its components being represented in the form of expansions in series in terms of the system of three types of the Hansen's vector functions (longitudinal and transverse ones), expressed via the scalar eigenfunctions. Such an approach presents a number of requirements, limiting the number of the systems in which this possibility of the Green's tensors creation exists, to the used systems of the orthogonal coordinates. We analyze these requirements briefly.

First of all, the form of the used coordinates system must provide separation of the variables of Helmhotz's three-dimentional equation in order to define the scalar eigenfunctions: $\varphi\,(\vec{q})$, $\psi\,(\vec{q})$ and $\chi\,(\vec{q})$. Such choice is defined by the Robertson's condition, and it limits 11 types of the coordinates systems. The scalar functions are represented as products of factor functions, each of which depends on only one variable, for example, $\varphi\,(\vec{q}) = \varphi_1\,(q_1)\,\varphi_2\,(q_2)\,\varphi_3\,(q_3)$ here.

However, if the variables are separated even in the scalar field equation in a concrete system of coordinates, then the variables cannot be separated yet in the Helmholtz's equations, and, even if they are separated in these equations, then the vector field components are so "shuffled," that each of them is included in all three equations. To avoid this the method of vector field separation into two parts is used [1.7]: one is obtained as "grad" of the scalar potential (the longitudinal component), the other, as "rot" of the vector potential (the transverse or solenoidal components). Then, one manages to create the complete system of the Hansen's vector eigenfunctions for the Helmholtz's vector equation in the following form on the basis of the three solutions of the Helmholtz's scalar equations $\varphi\,(\vec{q})$, $\psi\,(\vec{q})$ and $\chi\,(\vec{q})$ under the condition of the $h_1 = 1$ Lamé coefficient:

$$
\begin{aligned}
\vec{L} &= \operatorname{grad}\varphi\,(\vec{q})\,; \\
\vec{M} &= \operatorname{rot}\,(\vec{n}_1 w\psi\,(\vec{q}))\,; \\
\vec{N} &= \frac{1}{k}\operatorname{rotrot}\,(\vec{n}_1 w\chi\,(\vec{q}))\,,
\end{aligned}
\tag{1.34}
$$

where $\vec{n}_1$ is the unit vector for the q_1 coordinate, $1/k$ is the coefficient, included for that, so that the $\vec{L}$, $\vec{M}$, $\vec{N}$ dimensionality equal.

The requirements of equality to one of one of the Lamé coefficients and independence of ratio of another two from the coordinate corresponding to this coefficient, limit the number of "suitable" coordinates systems to six. They are: the rectangular coordinates, in which x, y or z can be chosen for q_1; three cylindrical systems of the coordinates, in which the z coordinate must be chosen as q_1; spherical and conical systems of coordinates, in which the r radius is as q_1 coordinate. $w = 1$ in the expressions (1.34) in the first four cases, $w = r-$ in the two last ones.

One should underscore, particularly, that the vector eigenfunctions system (1.34), created in such a way, can provide the boundary conditions performance for the vector fields only in a case where the boundary surfaces on which these conditions are set in the boundary problem coincide with the coordinate surfaces of the used coordinates system. That is, the Green's functions can be defined only for the spatial regions with the so called coordinate boundaries by the given method. We note here that this requirement defines the choice of the coordinates system in the boundary problems solution of different geometry.

The eigenfunctions of the Helmhotz's vector equation are structurally defined on the ground (1.34) in the form:

$$\vec{F}_n\,(\vec{q}) = \vec{L} + \vec{M} + \vec{N}, \tag{1.35}$$

where the four-valued index, designated via n conventionally here, can be defined for each $\vec{F}_n\,(\vec{q})$ eigenvector. The first number of this index shows which of these three systems $\vec{L}$, $\vec{M}$ or $\vec{N}$ the vector belongs to; the remaining three ones are the numbers of the $\varphi\,(\vec{q})$, $\psi\,(\vec{q})$ and $\chi\,(\vec{q})$ generating scalar eigenfunctions. It is possible to show that $\vec{F}_n\,(\vec{q})$ creates the complete orthogonal system of functions for all possible values of n, that is:

$$\iiint \vec{F}_n^*\,(\vec{q})\; \vec{F}_m\,(\vec{q})\;\mathrm{d}v = \begin{cases} 0 \text{ at } m \neq n \\ \lambda_n \text{ at } m = n \end{cases}, \tag{1.36}$$

where integration is performed over the whole spatial region; $\vec{F}_n^*\,(\vec{q})$ is the vector function, complex-conjugated to the $\vec{F}_n\,(\vec{q})$ function; λ_n is the norm of the corresponding eigenfunctions.

Using the operation of tensor multiplication of two vectors, the Green's tensor can be represented in the form of expansion in terms of the $\vec{F}_n\,(\vec{q})$ eigenvectors:

$$\hat{G}\,(\vec{q},\vec{q}') = 4\pi \sum_n \frac{\vec{F}_n^*\,(\vec{q}) \otimes \vec{F}_n\,(\vec{q}')}{\lambda_n\,(k_n^2 - k^2)}, \tag{1.37}$$

where $\vec{F}_n\,(\vec{q})$ satisfy the same boundary conditions as the components $\hat{G}\,(\vec{q},\vec{q}')$; k_n^2 are the corresponding eigenvalues; "$\otimes$" is the symbol of tensor multiplication. If $\vec{F}_n\,(\vec{q})$ is the complex vector, then the vector conjugated to it is included in the expansion (1.37) too, so $\hat{G}\,(\vec{q},\vec{q}')$ is the hermitian tensor. It is also seen, that $\hat{G}\,(\vec{q},\vec{q}')$ is the symmetrical tensor, and it has symmetry in respect to $\vec{q}$ and $\vec{q}'$.

We note, that the $\hat{G}\,(\vec{q},\vec{q}')$ tensor, considered as the function from k, has poles in all k_n eigenvalues, what's more, substraction equals to $2\pi\,\vec{F}_n^*\,(\vec{q})\,\vec{F}_n\,(\vec{q}')\big/(\lambda_n k_n)$ in the n-th pole. This corresponds to the infinite amplitude of the oscillations at the resonant frequency of constraining force (if damping is absent).

Let us create the Green's tensor for more interesting from the practical point of view system of the combined cylindrical coordinates (including rectangular ones) owing to the described method.

1.5.3 Tensor for the Region with Cylindrical Boundaries

To create the Green's tensor function of the Helmholtz's vector equation in the system of generalized cylindrical coordinates (including rectangular ones) we use the method considered above. Let (q_1, q_2, z) define the coordinates of the observation point here; (q_1', q_2', z') are the coordinates of the source point.

The tensor functions are located in the form of a series in the term of the vector functions system (1.34) on the ground of the representation (1.37). This expansion has the following form. For example, for $\hat{G}^E (\vec{q}, \vec{q}')$:

$$
\hat{G}^E (\vec{q}, \vec{q}') = \sum_{n=0}^{\infty} \sum_{m=0}^{\infty} \left\{ \frac{1}{\left(k_{nm}^E\right)^2 \lambda_{nm}^E} \left[\vec{z}^0, \nabla \psi_{nm} (q_1, q_2)\right] \otimes \left[\vec{z}^0, \nabla \psi_{nm}^* (q_1', q_2')\right] h_{nm}^E \right.
$$

$$
+ \frac{1}{\left(k_{nm}^M\right)^2 \lambda_{nm}^M} \nabla \chi_{nm} (q_1, q_2) \otimes \chi_{nm}^* (q_1', q_2') \, h_{nm}^M \qquad (1.38)
$$

$$
\left. + \frac{1}{\lambda_{nm}^M} \vec{z}^0 \chi_{nm} (q_1, q_2) \otimes \vec{z}^0 \chi_{nm}^* (q_1', q_2') \, h_{nm}^M \right\},
$$

where $\nabla u = \vec{q}_1^0 \frac{1}{h_1} \frac{\partial u}{\partial q_1} + \vec{q}_2^0 \frac{1}{h_2} \frac{\partial u}{\partial q_2}$; q_1, q_2 are the curvilinear coordinates in the cross section of the considered region with the h_1 and h_2 Lamé coefficients; $\vec{q}_1^0$ and $\vec{q}_2^0$ are the unit vectors; $\lambda_{nm}^{E(M)}$ are the eigenfunctions norms.

The $\psi_{nm} (q_1, q_2)$ and $\chi_{nm} (q_1, q_2)$ functions are mutually orthogonal by the eigenfunctions of the Helmholtz's two-dimensional equations:

$$
\begin{aligned}
\Delta \, \chi_{nm} (q_1, q_2) + \left(k_{nm}^M\right)^2 \chi_{nm} (q_1, q_2) = 0, \\
\Delta \, \psi_{nm} (q_1, q_2) + \left(k_{nm}^E\right)^2 \psi_{nm} (q_1, q_2) = 0
\end{aligned} \qquad (1.39)
$$

with the boundary conditions $\chi_{nm} (q_1, q_2) = \frac{\partial}{\partial n} \psi_{nm} (q_1, q_2) = 0$ in the expression (1.38).

The h_{nm}^E and h_{nm}^M functions (1.38) satisfy the inhomogeneous ordinary differential equations:

$$
\frac{\partial^2}{\partial z^2} h_{nm}^{E(M)} (z, z') + \left(\gamma_{nm}^{E(M)}\right)^2 h_{nm}^{E(M)} (z, z') = - 4\pi \delta \, (z - z'), \qquad (1.40)
$$

where $\gamma_{nm}^{E(M)} = \sqrt{k^2 - \left(k_{nm}^{E(M)}\right)^2}$, with the set boundary conditions on the ends of an interval on z.

We note that the form of the equations (1.40) for $h_{nm}^{E(M)} (z, z')$ and their solution do not depend upon the field distribution in the region cross section for the system of the cylindrical coordinates. They should be obtained according to the geometry of the boundary problem, set in the longitudinal direction.

The $\hat{G}^E (\vec{q}, \vec{q}')$ tensor function will have the following form for the regions in the considered coordinates system:

$$\hat{G}^E (\vec{q}, \vec{q}') = \begin{pmatrix} G_{11}^E (\vec{q}, \vec{q}') & G_{12}^E (\vec{q}, \vec{q}') & 0 \\ G_{21}^E (\vec{q}, \vec{q}') & G_{22}^E (\vec{q}, \vec{q}') & 0 \\ 0 & 0 & G_{33}^E (\vec{q}, \vec{q}') \end{pmatrix}, \qquad (1.41)$$

because we obtain five tensor components, differing from null, from the expression (1.38). We give the expression for $G_{33}^E (\vec{q}, \vec{q}')$ of the Green's tensor function component, for example, in an explicit form here because it will be used later in the book:

$$G_{33}^E (\vec{q}, \vec{q}') = G_{zz}^E (\vec{q}, \vec{q}') = \sum_{n=0}^{\infty} \sum_{m=0}^{\infty} \frac{1}{\lambda_{nm}^E} \chi_{nm} (q_1, q_2) \chi_{nm}^* (q_1, q_2) h_{nm}^E (z, z').$$
$$(1.42)$$

The Green's tensor functions of the $\hat{G}^M (\vec{q}, \vec{q}')$ magnetic type are also defined by the expression (1.38) at the following mutual changes: $\chi_{nm} (q_1, q_2) \Leftrightarrow \psi_{nm} (q_1, q_2); h_{nm}^E (z, z') \Leftrightarrow h_{nm}^M (z, z'); k_{nm}^E \Leftrightarrow k_{nm}^M; \lambda_{nm}^E \Leftrightarrow \lambda_{nm}^M$.

We note that the eigenfunctions of the transverse coordinates, their norms and eigenvalues for a great number of different types of the regions, the boundaries of which are formed by combinations of the coordinate surfaces of the system of the generalized cylindrical coordinates, have been obtained on the basis of the (1.39) equations solution in the monograph [1.8].

The Green's tensor functions of a magnetic type for a rectangular cross section of the waveguide, for which $G_{12}^{E(M)} (\vec{q}, \vec{q}') = G_{21}^{E(M)} (\vec{q}, \vec{q}') = 0$, are given in Appendix A. The Green's function for an infinite rectangular waveguide, for a half-infinite rectangular waveguide, for a rectangular resonator, for a half-infinite rectangular waveguide with the impedance end-face, and for a half-space over an infinite perfectly conducting plane are also represented in an explicit form in this appendix. Let us note that the $\hat{G}_{33}^{E(M)} (\vec{q}, \vec{q}')$ Green's functions for the waveguide with longitudinal–inhomogeneous dielectric loading have been obtained in an explicit form in [1.3].

1.5.4 Formulation of the Boundary Problem for the Coupling Hole in the Infinite Perfectly Conducting Screen

The boundary problem for the coupling hole, cut through in any perfectly conducting surface, is usually formulated in the following way. The slot aperture is conditionally "metallized," which reestablishes boundary surface homogeneity for both electromagnetic volumes, coupled via the slot-hole. We note that one can already speak about the Green's functions' use of these conjugated volumes under this condition. Having set the equivalent magnetic currents on the surface of the slot aperture, the electromagnetic currents are obtained with the help of the Green's functions in each volume. Using the boundary condition (1.17) further, the magnetic fields on the slot aperture, taking into consideration the original field of excitation are laced. Thus the integral-differential equation is obtained relative to the equivalent magnetic currents on the slot aperture.

However, formulation of the boundary problem for the coupling hole of two half-spaces can be obtained directly from Maxwell's equations, written in the integral representation [1.9] in the case of a thin infinite perfectly conducting screen. We analyze this possibility here because it represents fundamental interest in the electrodynamic theory of the coupling slot-holes.

Let us first consider the equivalency of mutual integral equations of macroscopic electrodynamics [1.9] and the Kirchhoff–Kotler's integral equations.

Let us put into consideration two $\vec{A}$ and $\vec{B}$ auxiliary vectors by the ratios:

$$\vec{A} = ik \int_V \varepsilon \vec{E}(\vec{r}') f\left(|\vec{r} - \vec{r}'|\right) d\vec{r}' - \operatorname{rot} \int_V \vec{H}(\vec{r}') f\left(|\vec{r} - \vec{r}'|\right) d\vec{r}',$$

$$\vec{B} = ik \int_V \mu \vec{H}(\vec{r}') f\left(|\vec{r} - \vec{r}'|\right) d\vec{r}' + \operatorname{rot} \int_V \vec{E}(\vec{r}') f\left(|\vec{r} - \vec{r}'|\right) d\vec{r}', \quad (1.43)$$

where $f\left(|\vec{r} - \vec{r}'|\right)$ is the abbreviated notation of the function $\dfrac{e^{-ik|\vec{r} - \vec{r}'|}}{|\vec{r} - \vec{r}'|}$, V is the volume, bounding the body scatter with the ε and μ parameters ($\varepsilon_1 = \mu_1 = 1$).

As can be seen, it is possible to rewrite the integral equations from [1.9] with the help of these vectors in the form

$$\vec{E}(\vec{r}) = \vec{E}_0(\vec{r}) + \frac{1}{4\pi i k} \left\{ \left(\operatorname{grad} \operatorname{div} + k^2\right) \vec{A} - ik \operatorname{rot} \vec{B} \right\},$$

$$\vec{H}(\vec{r}) = \vec{H}_0(\vec{r}) + \frac{1}{4\pi i k} \left\{ \left(\operatorname{grad} \operatorname{div} + k^2\right) \vec{B} + ik \operatorname{rot} \vec{A} \right\},$$

where $\vec{E}_0(\vec{r})$ and $\vec{H}_0(\vec{r})$ are the impressed fields of excitation. It is taken into account here that the observation point is outside the scattering body, and that is why

$$\left(\Delta + k^2\right) \int_V \vec{E}(\vec{r}') f\left(|\vec{r} - \vec{r}'|\right) d\vec{r}' = 0,$$

and

$$\left(\Delta + k^2\right) \int_V \vec{H}(\vec{r}') f\left(|\vec{r} - \vec{r}'|\right) d\vec{r}' = 0.$$

On the other hand, we have

$$ik \int_V \varepsilon \vec{E}(\vec{r}') f\left(|\vec{r} - \vec{r}'|\right) d\vec{r}' - \operatorname{rot} \int_V \vec{H}(\vec{r}') f\left(|\vec{r} - \vec{r}'|\right) d\vec{r}'$$

$$= ik \int_V \varepsilon \vec{E}(\vec{r}') f\left(|\vec{r} - \vec{r}'|\right) d\vec{r}' - \int_V \left[\nabla_{\vec{r}} f\left(|\vec{r} - \vec{r}'|\right), \vec{H}(\vec{r}')\right] d\vec{r}',$$

because the $\vec{H}(\vec{r}')$ vector is constant relative to differentiation on the $\vec{r}$ variable. We use the known ratio

$$\nabla_{\vec{r}} f\left(\left|\vec{r} - \vec{r}'\right|\right) = -\nabla_{\vec{r}'} f\left(\left|\vec{r} - \vec{r}'\right|\right),$$

because the $\vec{r}$ and $\vec{r}'$ variables are contained in the function of the $f\left(\left|\vec{r} - \vec{r}'\right|\right)$ in the form of $\left|\vec{r} - \vec{r}'\right|$ difference. That is

$$\left[\nabla_{\vec{r}} f\left(\left|\vec{r} - \vec{r}'\right|\right), \vec{H}(\vec{r}')\right] = -\left[\nabla_{\vec{r}'} f\left(\left|\vec{r} - \vec{r}'\right|\right), \vec{H}(\vec{r}')\right]$$
$$= f\left(\left|\vec{r} - \vec{r}'\right|\right) \operatorname{rot}_{\vec{r}'} \vec{H}(\vec{r}') - \operatorname{rot}_{\vec{r}'}\left(f\left(\left|\vec{r} - \vec{r}'\right|\right) \vec{H}(\vec{r}')\right)$$

and that is why

$$ik \int_V \varepsilon \vec{E}(\vec{r}') f\left(\left|\vec{r} - \vec{r}'\right|\right) d\vec{r}' - \operatorname{rot} \int_V \vec{H}(\vec{r}') f\left(\left|\vec{r} - \vec{r}'\right|\right) d\vec{r}'$$
$$= ik \int_V \varepsilon \vec{E}(\vec{r}') f\left(\left|\vec{r} - \vec{r}'\right|\right) d\vec{r}' - \int_V f\left(\left|\vec{r} - \vec{r}'\right|\right) \operatorname{rot}_{\vec{r}'} \vec{H}(\vec{r}') d\vec{r}'$$
$$+ \int_V \operatorname{rot}_{\vec{r}'}\left(f\left(\left|\vec{r} - \vec{r}'\right|\right) \vec{H}(\vec{r}')\right) d\vec{r}'.$$

Because the fields satisfy Maxwell's equations (1.2) for all $\vec{r}'$ points inside the V volume

$$\operatorname{rot}\vec{H} = ik\varepsilon\vec{E}, \qquad \operatorname{rot}\vec{E} = -ik\mu\vec{H}$$

and due to the divergence theorem

$$\int_V \operatorname{rot}_{\vec{r}'}\left(f\left(\left|\vec{r} - \vec{r}'\right|\right) \vec{H}(\vec{r}')\right) d\vec{r}' = \oint_S f\left(\left|\vec{r} - \vec{r}'\right|\right)\left[\vec{n}, \vec{H}(\vec{r}')\right] ds',$$

we finally have

$$\vec{A} = \oint_S f\left(\left|\vec{r} - \vec{r}'\right|\right)\left[\vec{n}, \vec{H}(\vec{r}')\right] ds'. \tag{1.44}$$

Similar calculations also give the ratio

$$\vec{B} = -\oint_S f\left(\left|\vec{r} - \vec{r}'\right|\right)\left[\vec{n}, \vec{E}(\vec{r}')\right] ds'. \tag{1.45}$$

Thus, the integral equations of the macroscopic electrodynamics can be written in the form of

$$\vec{E}(\vec{r}) = \vec{E}_0(\vec{r}) + \frac{1}{4\pi i k} \left\{ \left(\text{grad div} + k^2 \right) \oint_S f\left(|\vec{r} - \vec{r}'|\right) \left[\vec{n}, \vec{H}(\vec{r}')\right] ds' \right.$$

$$\left. +ik \text{ rot} \oint_S f\left(|\vec{r} - \vec{r}'|\right) \left[\vec{n}, \vec{E}(\vec{r}')\right] ds' \right\}, \qquad (1.46)$$

$$\vec{H}(\vec{r}) = \vec{H}_0(\vec{r}) + \frac{1}{4\pi i k} \left\{ -\left(\text{grad div} + k^2 \right) \oint_S f\left(|\vec{r} - \vec{r}'|\right) \left[\vec{n}, \vec{E}(\vec{r}')\right] ds' \right.$$

$$\left. +ik \text{ rot} \oint_S f\left(|\vec{r} - \vec{r}'|\right) \left[\vec{n}, \vec{H}(\vec{r}')\right] ds' \right\},$$

where integration is made along the whole S surface, which covers the scattering body. It is seen from the given ratios, that the scattering field is completely defined by tangential components of the complete internal field on the whole surface of the scattering body. The dielectric constant of the inner medium is already absent, and limiting process to the perfectly conducting body is performed with the help of the boundary condition (1.19). The equations (1.46) are called the Kirchhoff–Kotler's equations.

The perfectly conducting screen is a limit case of the thin but high-conductive body of arbitrary geometry in two (x and y) dimensions, when the screen thickness tends to zero, and its electro-conductivity to infinity, so that the doubled thickness of the skin-layer always stays less than the screen thickness. Thus, the surface currents are excited on the screen surface; what is more, the current on one side of the screen can influence the current on the other side of the screen only via space, but not its thickness. Consequently, the Kirchhoff–Kotler's equations are more fit to consider wave diffraction problems on the infinitely thin perfectly conducting metallic screens.

Let us consider the limit case, when the screen of finite thickness stretches along the OX, OY axes up to infinity in the XOY plane and creates the infinitely thin perfectly conducting plane of $z = 0$.

The integral equations of the electromagnetic waves diffraction are obtained directly from Kirchhoff–Kotler's equations by integration along the $z + 0$ surface of this plane for the $z > 0$ half-space and along $z - 0$ surface for the $z < 0$ half-space on such a plane, that is

$$\vec{E}(\vec{r}) = \vec{E}_0(\vec{r}) + \frac{1}{4\pi ik} \left\{ \left(\text{grad div} + k^2 \right) \oint_S \left[\vec{n}^{\pm}, \vec{H} \right] \frac{e^{-ik|\vec{r}-\vec{r}'|}}{|\vec{r}-\vec{r}'|} ds' \right.$$

$$\left. + ik \, \text{rot} \oint_S \left[\vec{n}^{\pm}, \vec{E} \right] \frac{e^{-ik|\vec{r}-\vec{r}'|}}{|\vec{r}-\vec{r}'|} ds' \right\},$$

$$\vec{H}(\vec{r}) = \vec{H}_0(\vec{r}) + \frac{1}{4\pi ik} \left\{ -\left(\text{grad div} + k^2 \right) \oint_S \left[\vec{n}^{\pm}, \vec{E} \right] \frac{e^{-ik|\vec{r}-\vec{r}'|}}{|\vec{r}-\vec{r}'|} ds' \right.$$

$$\left. + ik \, \text{rot} \oint_S \left[\vec{n}^{\pm}, \vec{H} \right] \frac{e^{-ik|\vec{r}-\vec{r}'|}}{|\vec{r}-\vec{r}'|} ds' \right\},$$

where integration is made along the whole plane, and it is taken into account that the outer normal is positive on the $z + 0$ surface and negative on the $z - 0$ surface in each point of the surface in a general case.

Now we take into consideration that the plane is perfectly conducting, that is $\left[\vec{n}, \vec{E} \right] = 0$ on its surface, and the electromagnetic field does not penetrate outside the plane. If the external wave falls out of the region of $z < 0$, then it stays as a reflected wave in this region, that is at $z < 0$

$$\vec{E}(\vec{r}) = \vec{E}_0(\vec{r}) + \frac{1}{4\pi ik} \left\{ -\left(\text{grad div} + k^2 \right) \int_S \left[\vec{n}, \vec{H} \right] \frac{e^{-ik|\vec{r}-\vec{r}'|}}{|\vec{r}-\vec{r}'|} ds' \right\},$$

$$\vec{H}(\vec{r}) = \vec{H}_0(\vec{r}) - \frac{1}{4\pi ik} \left\{ ik \, \text{rot} \int_S \left[\vec{n}, \vec{H} \right] \frac{e^{-ik|\vec{r}-\vec{r}'|}}{|\vec{r}-\vec{r}'|} ds' \right\},$$

(1.47)

then at $z > 0$

$$\vec{E}(\vec{r}) = \vec{E}_0(\vec{r}) + \frac{1}{4\pi ik} \left\{ \left(\text{grad div} + k^2 \right) \int_S \left[\vec{n}, \vec{H} \right] \frac{e^{-ik|\vec{r}-\vec{r}'|}}{|\vec{r}-\vec{r}'|} ds' \right\} = 0,$$

$$\vec{H}(\vec{r}) = \vec{H}_0(\vec{r}) + \frac{1}{4\pi ik} \left\{ ik \, \text{rot} \int_S \left[\vec{n}, \vec{H} \right] \frac{e^{-ik|\vec{r}-\vec{r}'|}}{|\vec{r}-\vec{r}'|} ds' \right\} = 0.$$

(1.48)

Consequently, the known reflection of the electromagnetic field is formed over the perfectly conducting surface in $z < 0$ region, and the field is absent in the $z > 0$ region.

Let us consider now the case where a slot with the σ square appears in the $z = 0$ continuous plane. We consider that this slot's linear sizes are small in comparison with the wavelength. The electromagnetic field will radiate to the $z > 0$ space via this slot.

The corresponding integral equations for the $z > 0$ region will now have the form

$$\vec{E}(\vec{r}) = \vec{E}_0(\vec{r}) + \frac{1}{4\pi i k} \left\{ \left(\text{grad div} + k^2 \right) \int\limits_{S-\sigma} \left[\vec{n}, \vec{H} \right] \frac{e^{-ik|\vec{r}-\vec{r}'|}}{|\vec{r} - \vec{r}'|} ds' \right\} \qquad (1.49)$$
$$+ \frac{1}{4\pi} \text{rot} \int\limits_{\sigma} \left[\vec{n}, \vec{E} \right] \frac{e^{-ik|\vec{r}-\vec{r}'|}}{|\vec{r} - \vec{r}'|} ds',$$

$$\vec{H}(\vec{r}) = \vec{H}_0(\vec{r}) + \frac{1}{4\pi} \left\{ \text{rot} \int\limits_{S-\sigma} \left[\vec{n}, \vec{H} \right] \frac{e^{-ik|\vec{r}-\vec{r}'|}}{|\vec{r} - \vec{r}'|} ds' \right\} \qquad (1.50)$$
$$+ \frac{1}{4\pi i k} \left\{ -\left(\text{grad div} + k^2 \right) \int\limits_{\sigma} \left[\vec{n}, \vec{E} \right] \frac{e^{-ik|\vec{r}-\vec{r}'|}}{|\vec{r} - \vec{r}'|} ds' \right\},$$

and because $S - \sigma \approx S$ at small sizes of the slot, we have expression for the slot radiated electromagnetic field in the form:

$$\vec{E}(\vec{r}) = \frac{1}{4\pi} \text{rot} \int\limits_{\sigma} \left[\vec{n}, \vec{E} \right] \frac{e^{-ik|\vec{r}-\vec{r}'|}}{|\vec{r} - \vec{r}'|} ds',$$
$$\vec{H}(\vec{r}) = -\frac{1}{4\pi i k} \left(\text{grad div} + k^2 \right) \int\limits_{\sigma} \left[\vec{n}, \vec{E} \right] \frac{e^{-ik|\vec{r}-\vec{r}'|}}{|\vec{r} - \vec{r}'|} ds', \qquad (1.51)$$

or having performed transition to the equivalent magnetic current $\vec{j}^M = \frac{c}{4\pi} \left[\vec{n}, \vec{E} \right]$ from $\left[\vec{n}, \vec{E} \right]$ in the equations (1.51), we have:

$$\vec{E}(\vec{r}) = \frac{1}{c} \text{rot} \int\limits_{\sigma} \vec{j}^M(\vec{r}') \frac{e^{-ik|\vec{r}-\vec{r}'|}}{|\vec{r} - \vec{r}'|} ds',$$
$$\vec{H}(\vec{r}) = -\frac{1}{i\omega} \left(\text{grad div} + k^2 \right) \int\limits_{\sigma} \vec{j}^M(\vec{r}') \frac{e^{-ik|\vec{r}-\vec{r}'|}}{|\vec{r} - \vec{r}'|} ds'. \qquad (1.52)$$

References

1.1. Debye, P.: Lichtuck auf Kugelu von beliebigen materiab. Ibid. 57–136 (1909)
1.2. Weinstein, L.A.: Electromagnetic Waves. Radio & Communication, Moskow (1988) (in Russian)
1.3. Penkin, Yu.M., Katrich, V.A.: Excitation of Electromagnetic Waves in the Volumes with Coordinate Boundaries. Fakt, Kharkov (2003) (in Russian)

1.4. Mittra, R. (ed.): Computer Techniques for Electromagnetics. Pergamon Press, Oxford, New York, (1973)

1.5. Popovich, B.: Polynomial approximation of current along thin symmetrical cylindrical dipoles. Proc. Inst. Elec. Eng. **117**, 873–878 (1970)

1.6. King, R.W.P., Mack, R.B., Sandler S.S.: Arrays of Cylindrical Dipoles. Cambridge University Press, NY (1968)

1.7. Morse, P.M., Feshbach, H.: Methods of Theoretical Physics. McGraw-Hill, New York (1953)

1.8. Felsen, L.B., Marcuvitz, N.: Radiation and Scattering of Waves. Prentice-Hall, New Jersey, (1973)

1.9. Khizhnyak, N.A.: Integral Equations of Macroscopical Electrodynamics. Naukova dumka, Kiev (1986) (in Russian)

Chapter 2
Problem Formulation and Initial Integral Equations; Averaging Method

The problem of electromagnetic coupling of two waveguides through apertures in their common walls is a classical problem that attracted the attention of many investigators, starting with the paper written by Bethe in 1944 [2.1]. The narrow slots with the length of $2L$ commensurable with the operating wavelength of λ have especially been studied [2.2–2.10] The investigations of the slots located both in broad and narrow walls of rectangular waveguides have been conducted by different methods, namely: analytical [2.2, 2.3], variational [2.4, 2.5], and numerical [2.6–2.8], among which the most effective methods are the moments method and its particular case known as Galerkin's method, the equivalent circuits method [2.9, 2.10], and also the finite elements method and the moments method [2.11].

At present, commercial finite elements software (e.g., "Ansoft's HFSS and Designer," "CST Microwave Studio," "Zeland," and others) is available to solve such problems. However, these programs need intensive memory, and sometimes they are very slow, for example, in electrically long slots and multi-slot systems analysis.

On the other hand, the approximate methods, mentioned above, have some drawbacks. The known analytical solutions have a limited range of applicability ($kL \cong \pi/2$, where $k = 2\pi/\lambda$) and the variational and equivalent circuits methods suppose the presence of a priori information about the distribution function of the equivalent slot magnetic current. Even an approximation of this information is unknown (for example, for electrically longitudinal slots) in some cases. That is why there exists a need to develop approximate methods that provide fast, sufficiently accurate calculations of simple waveguide-slot structures.

2.1 General Problem Formulation and Transition to the Case of a Narrow Slot

Let two volumes (limited by ideally conducting flat surfaces) be coupled between each other by the slot in the common unlimited thin wall. Using the boundary condition (1.18) of the tangential magnetic field continuity on the S_{sl} surface of

M.V. Nesterenko et al., *Analytical and Hybrid Methods in the Theory of Slot-Hole Coupling of Electrodynamic Volumes,* DOI: 10.1007/978-0-387-76362-0_2,
© Springer Science+Business Media, LLC, 2008

the coupling aperture, we obtain the following integral equation concerning the equivalent magnetic current [2.12]:

$$(\text{graddiv} + k^2) \int_{S_{sl}} \hat{G}_m^\Sigma(\vec{r}, \vec{r}') \vec{J}^m(\vec{r}') d\vec{r}' = -i\omega \vec{H}_0^\Sigma(\vec{r}). \tag{2.1}$$

Here: $\vec{r}$ is the observation point radius-vector; $\vec{r}'$ is the source radius-vector; $\vec{J}^m(\vec{r})$ is the magnetic current surface density on the aperture; $\hat{G}_m^\Sigma(\vec{r}, \vec{r}') = \hat{G}_m^e(\vec{r}, \vec{r}') + \hat{G}_m^i(\vec{r}, \vec{r}')$, $\hat{G}_m^{e,i}(\vec{r}, \vec{r}')$ are the magnetic dyadic Green's functions; $\vec{H}_0^\Sigma(\vec{r}) = \vec{H}_0^i(\vec{r}) - \vec{H}_0^e(\vec{r})$ are the impressed sources fields in the internal (index "i" for the region 1 and 2) and the external (index "e" for regions 3 and 4) volumes.

The $\hat{G}_m^{e,i}(\vec{r}, \vec{r}')$ functions are the following [1.9]:

$$\hat{G}_m^{e,i}(\vec{r}, \vec{r}') = \hat{I} G(\vec{r}, \vec{r}') + \hat{G}_{0m}^{e,i}(\vec{r}, \vec{r}'), \tag{2.2}$$

where $\hat{I}$ is the unit dyadic, $G(\vec{r}, \vec{r}') = \frac{e^{-ik|\vec{r}-\vec{r}'|}}{|\vec{r}-\vec{r}'|}$ is the Green's free space function and $\hat{G}_{0m}^{e,i}(\vec{r}, \vec{r}')$ are the regular everywhere dyadic functions providing satisfaction of boundary conditions for $\hat{G}_m^{e,i}(\vec{r}, \vec{r}')$ functions on the internal surface of the couplesd volumes.

Equation (2.1) is rather difficult to analyze in a general case; however, for the narrow slots ($d/_{2L} <<1$, $d/_\lambda <<1$), where d is the slot width, the equation is sufficiently simplified. In this case the slot current can be written in the following way (index "m" is omitted):

$$\vec{J}(\vec{r}) = \vec{e}_s J(s) \chi(\xi), \tag{2.3}$$

$$J(\pm L) = 0, \quad \int_\xi \chi(\xi) d\xi = 1,$$

where s and ξ are the longitudinal and transverse local slot coordinates (Fig. 2.1); $\vec{e}_s$ is the unit vector; $\chi(\xi)$ is the set function accounting for the peculiarities of the electrostatic field on the slot edge [2.13]:

$$\chi(\xi) = \frac{1/\pi}{\sqrt{(d/2)^2 - \xi^2}}. \tag{2.4}$$

Thus, the $\vec{J}(\vec{r})$ current problem in the $\vec{H}_0^\Sigma(\vec{r})$ field reduces to the determination of the $J(s)$ current distribution function.

Let us consider the slot to be rectilinear and the impressed field in the external volume to be absent, that is $\vec{H}_o^e(\vec{r}) = 0$. Then substituting (2.3) and (2.4) into (2.1) we get:

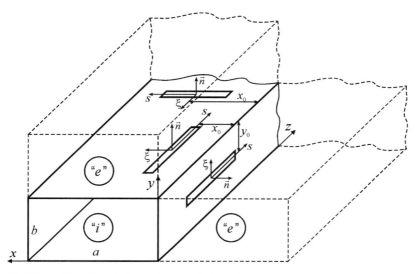

Fig. 2.1 The problem formulation and the symbols used

$$\left(\frac{d^2}{ds^2} + k^2\right) \int_{-L}^{L} J(s')[G_s^e(s, s') + G_s^i(s, s')]ds' = -i\omega H_{0s}^i(s), \qquad (2.5)$$

$$G_s^{e,i}(s, s') = 2\frac{e^{-ikR(s,s')}}{R(s, s')} + G_{0s}^{e,i}(s, s'). \quad R(s, s') = \sqrt{(s - s')^2 + (d/4)^2}. \quad (2.6)$$

Here we take into account the fact that for the sources on the flat surface

$$\hat{G}^{e,i}(s, \xi; s', \xi') = 2\hat{I}G(s, \xi; s', \xi') + \hat{G}_0^{e,i}(s, \xi; s', \xi')$$

we have

$$G(s, s') = \int_{\xi} G(s, \xi; s', \xi')d\xi', \quad \hat{G}_0^{e,i}(s, s') = \int_{\xi} \hat{G}_0^{e,i}(s, \xi; s', \xi')\chi(\xi')d\xi'.$$

It must be noted that in the kernel (2.6) of the integral equation (2.5) the approximate expression for $|\vec{r} - \vec{r}'|$ in the transverse coordinate dependence is chosen in the form of $(\xi - \xi')^2 \cong (d/4)^2$, as it is usually used in the vibrator antenna theory [2.14], and it is precisely the form for the slots on metallic surfaces [2.12, 2.13, 2.15].

Isolating the logarithmic singularity in the equation (2.5) analogically with [2.2, 2.12, 2.14] we obtain:

$$\int_{-L}^{L} J(s')\frac{e^{-ikR(s,s')}}{R(s,s')}ds' = J(s)\Omega(s) + \int_{-L}^{L}\frac{J(s')e^{-ikR(s,s')} - J(s)}{R(s,s')}ds', \qquad (2.7)$$

where $\Omega(s) = \int_{-L}^{L}\frac{ds'}{R(s,s')}$. Suppose due to [2.2] $\Omega(s) \approx \Omega(0) = 2\ln\frac{8L}{d}$, we obtain the integral–differential equation with a small parameter:

$$\frac{d^2 J(s)}{ds^2} + k^2 J(s) = \alpha\{i\omega H_{0s}(s) + F[s, J(s)] + F_0[s, J(s)]\}, \qquad (2.8)$$

where $\alpha = \frac{1}{8\ln[d/(8L)]}$ is the natural small ($|\alpha| << 1$) parameter of the problem; $H_{0s}(s)$ is the component of the field of the impressed sources on the slot axis;

$$F[s, J(s)] = 4\left[-\frac{dJ(s')}{ds'}\frac{e^{-ikR(s,s')}}{R(s,s')}\bigg|_{-L}^{L} + \left(\frac{2dJ(s)}{ds} + J(s)\frac{d}{ds}\right)\frac{1}{R(s,s')}\right]$$

$$+4\int_{-L}^{L}\frac{\left[\frac{d^2 J(s')}{ds'^2} + k^2 J(s')\right]e^{-ikR(s,s')} - \left[\frac{d^2 J(s)}{ds^2} + k^2 J(s)\right]}{R(s,s')}ds'$$

$$\qquad (2.9)$$

is the slot own field in the infinite screen;

$$F_0[s, J(s)] = -\frac{dJ(s')}{ds'}[G_{0s}^e(s, s') + G_{0s}^i(s, s')]\bigg|_{-L}^{L}$$

$$+\int_{-L}^{L}\left[\frac{d^2 J(s')}{ds'^2} + k^2 J(s')\right][G_{0s}^e(s, s') + G_{0s}^i(s, s')]ds' \quad (2.10)$$

is the slot own field repeatedly reflected from the coupled volumes' walls.

We must note that the inequality $|\alpha| <<1$ is valid in sufficiently wide limits of the ratio $\left(\frac{d}{2L}\right)$ variation: for example, for $\left(\frac{d}{2L}\right) = 0.1$ we have $|\alpha| = 0.034$, and for $\left(\frac{d}{2L}\right) = 0.3$ we have $|\alpha| = 0.048$.

2.2 Approximate Analytical Methods of Integral Equations Solutions for the Current

Rigorous equations solution for the slot magnetic current in the closed form cannot be obtained; however, it is not inconsequential that it is impossible to approximate it sufficiently accurately by approximate solutions. To define the advantages and disadvantages of one or another of the earlier known analytical methods let us find the equation solution (2.5) by the expansion method of the searched function into the series at the small parameter and by the consistent iterations method (the iterations method).

2.2.1 Small Parameter Expansion Method

Let us rewrite equation (2.5) in the following form, taking into account (2.6) (the "i" index is omitted):

$$\left(\frac{d^2}{ds^2} + k^2\right) \int_{-L}^{L} J(s')4\frac{e^{-ikR(s,s')}}{R(s,s')} ds' = -i\omega H_{0s}(s) - f_0[s, J(s)], \quad (2.11)$$

where

$$f_0[s, J(s)] = \left(\frac{d^2}{ds^2} + k^2\right) \int_{-L}^{L} J(s')G_{0s}^{\Sigma}(s, s')ds', \quad (2.12)$$

$$G_{0s}^{\Sigma}(s, s') = G_{0s}^{e}(s, s') + G_{0s}^{i}(s, s').$$

Exchanging the ds' differential on dR in (2.11) and taking into account that

$$\left.\begin{array}{l} s' = s - \sqrt{R^2 - (d/4)^2}, s' \leq s \\ s' = s + \sqrt{R^2 - (d/4)^2}, s' \geq s \end{array}\right\},$$

we transform the equation (2.11) into the form:

$$\left(\frac{d^2}{ds^2} + k^2\right) 4\left\{-\int_{-L}^{s} J(s')e^{-ikR}d\ln[C(R + \sqrt{R^2 - (d/4)^2})]\right.$$

$$\left. + \int_{s}^{L} J(s')e^{-ikR}d\ln[C(R + \sqrt{R^2 - (d/4)^2})]\right\} = -i\omega H_{0s}(s) - f_0[s, J(s)],$$

$$(2.13)$$

where C is the arbitrary constant.

Making integration by sides with the current boundary conditions usage (2.3) $J(\pm L) = 0$ in (2.13), we obtain:

$$\left(\frac{d^2}{ds^2} + k^2\right) 4\left\{2J(s)e^{-ikd/4}\ln C\frac{d}{4} + \int_{-L}^{s} \ln[C(R + \sqrt{R^2 - (d/4)^2})]\,d[J(s')e^{-ikR}]\right.$$

$$\left. - \int_{s}^{L} \ln[C(R + \sqrt{R^2 - (d/4)^2})]\,d[J(s')e^{-ikR}]\right\} = i\omega H_{0s}(s) + f_0[s, J(s)].$$

$$(2.14)$$

Supposing $e^{-ikd/4} \cong 1$, and if we choose $C = 1/2L$, we reduce the equation (2.11) to the following integral–differential equation for the current with the small parameter:

$$\frac{d^2 J(s)}{ds^2} + k^2 J(s) = \alpha \{i\omega H_{0s}(s) + f[s, J(s)] + f_0[s, J(s)]\}. \qquad (2.15)$$

Here $\alpha = \frac{1}{8\ln[d/(8L)]}$ is the small parameter,

$$f[s, J(s)] = -\left(\frac{d^2}{ds^2} + k^2\right) 4 \int\limits_{-L}^{L} \text{sign}(s - s') \ln \frac{R + (s - s')}{2L} \frac{d}{ds'} \left[J(s')e^{-ikR}\right] ds' \qquad (2.16)$$

is the slot own field in the infinite screen.

Let us note that equation (2.15) and equation (2.8) are analogically alike, but the slot own field functions differ from each other sufficiently in these equations because these equations are obtained by different methods.

Let us represent $J(s)$ in the form of the power series at the α small parameter:

$$J(s) = J_0(s) + \alpha J_1(s) + \alpha^2 J_2(s) + \ldots \qquad (2.17)$$

Substitution of (2.17) into (2.12) and (2.16) allows us to expand into the analogous series

$$f_\Sigma[s, J(s)] = f_\Sigma[s, J_0(s)] + \alpha \, f_\Sigma[s, J_1(s)] + \alpha^2 f_\Sigma[s, J_2(s)] + \ldots, \qquad (2.18)$$

where $f_\Sigma[s, J(s)] = f[s, J(s)] + f_0[s, J(s)]$ is the total slot own field.

Having substituted (2.17) and (2.18) into equation (2.15) and having equated the multipliers at the α similar degrees in both the right and left equation parts between each other, we obtain the following differential equations system:

$$\frac{d^2 J_0(s)}{ds^2} + k^2 J_0(s) = 0,$$

$$\frac{d^2 J_1(s)}{ds^2} + k^2 J_1(s) = i\omega H_{0s}(s) + f_\Sigma[s, J_0(s)],$$

$$\frac{d^2 J_2(s)}{ds^2} + k^2 J_2(s) = f_\Sigma[s, J_1(s)], \qquad (2.19)$$

$$\cdots\cdots\cdots\cdots\cdots\cdots\cdots\cdots\cdots\cdots\cdots\cdots\cdots\cdots\cdots$$

$$\frac{d^2 J_n(s)}{ds^2} + k^2 J_n(s) = f_\Sigma[s, J_{n-1}(s)],$$

where the solution of each of the equations is searched at the boundary conditions of the form (2.3), namely: $J_0(\pm L) = 0$, $J_1(\pm L) = 0$, $J_2(\pm L) = 0,\ldots,J_n(\pm L) = 0$.

The first equation of the system (2.19) has the solution, independent of the $H_{0s}(s)$ exciting field:

$$J_0(s) = C_1 \cos ks + C_2 \sin ks, \tag{2.20}$$

which satisfies the boundary conditions only when the ratio is performed

$$C_1 = 0, \ 2L = m\lambda, \quad C_2 = 0, \ 2L = (2n+1)\frac{\lambda}{2}, \tag{2.21}$$

where m and n are the integers.

The current equals at the $2L$ slot length, unsatisfying the conditions (2.21), $J_0 \equiv 0$, $f_\Sigma[s, J_0(s)] \equiv 0$ in the first approximation:

$$J(s) = \alpha J_1(s) = -\alpha \frac{i\omega/k}{\sin 2kL} \left\{ \sin k(L-s) \int_{-L}^{s} H_{0s}(s') \sin k(L+s')\,ds' \right.$$

$$\left. + \sin k(L+s) \int_{s}^{L} H_{0s}(s') \sin k(L-s')\,ds' \right\}. \tag{2.22}$$

As it is seen, the slot own field functions $f[s, J(s)]$ are not involved in the current expression, which, mainly defines the slot resonant and energetic characteristics. Evidently, it is necessary to obtain the following approximations to take into account $f_\Sigma[s, J(s)]$, which, however, meets sufficient mathematical difficulties.

The impressed field has two components —the symmetrical (index "s") and anti-symmetrical (index "a") along the slot respectively to its center for the longitudinal slot in the common broad wall of standard rectangular waveguides, one of which is exciting by the wave of the H_{10} main type:

$$H_{0s}(s) = H_{0s}^s(s) + H_{0s}^a(s) = H_0 \cos \frac{\pi x_0}{a}(\cos \gamma s - i \sin \gamma s). \tag{2.23}$$

Here H_0 is the amplitude of the H_{10} wave type, incident from the $z = -\infty$ region, x_0 is the distance from the narrow wall of a waveguide to the slot axial line (Fig. 2.1), $\gamma = \sqrt{k^2 - (\pi/a)^2}$. The induced current in the slot will equal due to (2.22) to:

$$J(s) = -\alpha \frac{i\omega H_0 \cos \frac{\pi x_0}{a}}{(\pi/a)^2} \left\{ \frac{(\cos ks \cos \gamma L - \cos kL \cos \gamma s)}{\cos kL} \right.$$
$$\left. -i\frac{(\sin ks \sin \gamma L - \sin kL \sin \gamma s)}{\sin kL} \right\}. \tag{2.24}$$

For the transverse slot in the broad wall of a waveguide

$$H_{0s}(s) = H_0 \sin \frac{\pi}{a}(x_0 + s), \qquad (2.25)$$

where x_0 is the slot center shift from the narrow wall of a waveguide, and the expression for the current in the first approximation at $x_0 = a/2$ has the form:

$$J(s) = -\alpha H_0 \frac{i\omega \left(\cos ks \cos \frac{\pi}{a}L - \cos kL \cos \frac{\pi}{a}s \right)}{\gamma^2 \cos kL}. \qquad (2.26)$$

Let us note in conclusion that criterion of applicability of the formulas (2.24) and (2.26) is the condition

$$\left| kL - n\frac{\pi}{2} \right| >> |\alpha|, \qquad (2.27)$$

which significantly limits the possibility of using the obtained solution in practice together with the ratios (2.21).

2.2.2 Method of the Consistent Iterations

To eliminate the above-mentioned disadvantages of the current integral equation solution by the small parameter expansion method, let us use the method of consistent iterations [2.14], widely used to investigate the characteristics of thin vibrators. Inverting the differential operator on the left side (2.5), we obtain the following integral equation:

$$\int_{-L}^{L} J(s')G_s^{\Sigma}(s, s') \, ds' = C_1 \cos ks + C_2 \sin ks - \frac{i\omega}{k} \int_{-L}^{s} H_{0s}(s') \sin k(s - s') \, ds',$$

$$(2.28)$$

where it is taken into account that $G_s^{\Sigma}(s, s') = G_s^e(s, s') + G_s^i(s, s')$.

To define one of the arbitrary constants C_1 and C_2 it is necessary to use the conditions of the symmetry [2.14], which are explicitly connected with the slot mode of excitation. In other words, it is necessary to give concrete expression to the field of the impressed sources $H_{0s}(s)$ at the given stage of the equation solution for the slot magnetic field by the iterations method. Suppose $H_{0s}(s) = H_0 \cos \frac{\pi}{a}s$, which corresponds to the symmetrical transverse slot in the broad wall of the rectangular waveguide. In this case

$$\int_{-L}^{L} J(s')G_s^{\Sigma}(s, s')\, ds' = C_1 \cos ks + \frac{i\omega}{\gamma^2} H_0[\cos ks\ f(kL) - \cos \frac{\pi}{a} s],$$

$$f(kL) = \cos kL \cos \frac{\pi}{a} L + \frac{\pi}{ka} \sin kL \sin \frac{\pi}{a} L. \tag{2.29}$$

We further transform equation (2.29), isolating the logarithmic singularity in the kernel of the equation similarly with (2.7)

$$J(s) = -\alpha \left\{ C_1 \cos ks + \frac{i\omega}{\gamma^2} H_0[\cos ks\ f(kL) - \cos \frac{\pi}{a} s] \right\}$$

$$+ \alpha \int_{-L}^{L} \left[J(s')G_s^{\Sigma}(s, s') - \frac{4J(s)}{R(s, s')} \right] ds', \tag{2.30}$$

where $\alpha = \frac{1}{8 \ln[d/(8L)]}$ is the small parameter, coinciding with the one obtained earlier in Subsection 2.2.1. Further, due to the method described in [2.14], let us suppose in (2.30) $s = L$ and subtract the obtained expression from (2.30) (in fact, we subtract 0, because $J(L) \equiv 0$). So equation (2.30) transits into the following expression:

$$J(s) = -\alpha \left\{ C_1(\cos ks - \cos kL) + \frac{i\omega}{\gamma^2} H_0 \right.$$

$$\times \left[(\cos ks - \cos kL) f(kL) - \left(\cos \frac{\pi}{a} s - \cos \frac{\pi}{a} L \right) \right] \Big\}$$

$$+ \alpha \left\{ \int_{-L}^{L} \left[J(s')G_s^{\Sigma}(s, s') - \frac{J(s)}{R(s, s')} \right] ds' - \int_{-L}^{L} J(s')G_s^{\Sigma}(L, s')\, ds' \right\}. \tag{2.31}$$

If we choose the item in the right side of the first line of the expression (2.31) for the $J_0(s)$ current as the zeroth approximation and use condition (2.3) to define the C_1 constant, we obtain:

$$J_0(s) = -\alpha H_0 \frac{i\omega \left(\cos ks \cos \frac{\pi}{a} L - \cos kL \cos \frac{\pi}{a} s \right)}{\gamma^2 \cos kL}, \tag{2.32}$$

which identically coincides with the expression (2.26), obtained by the small parameter expansion method in the first approximation.

Substituting now (2.32) into (2.31) with accuracy to the terms of the α^2 order, we obtain the first approximation for the current:

$$J_1(s) = -\alpha H_0 \frac{i\omega \left(\cos ks \cos \frac{\pi}{a} L - \cos kL \cos \frac{\pi}{a} s \right)}{\gamma^2 [\cos kL + \alpha F(kd, kL)]}, \tag{2.33}$$

where

$$F(kd, kL) = \int\limits_{-L}^{L} [(\cos ks' - \cos kL)G_s^{\Sigma}(L, s')] \, ds' \qquad (2.34)$$

is the slot own field function, giving the opportunity to analyze both adjusted ($\cos kL = 0$) and unadjusted ($\cos kL \neq 0$) slots with the help of the formula in the first approximation at α (unlike the small parameter expansion method).

Let us note that Stevenson obtained the current distribution in an arbitrary oriented slot in the broad wall of a rectangular waveguide in the first approximation by the iterations method in [2.2]; however, Stevenson supposed $kL = \pi/2$ both in the function of the current distribution and in the slot own field function.

Thus the solution of the quasi-one-dimensional equation for the magnetic current in the slot-holes of coupling of the electrodynamic volumes by the small parameter expansion method leads to different expressions for the current in the case of adjusted (the frequency of the impressed field slightly differs from the slot eigen frequency) and unadjusted slots (when this condition is not performed), though the solution can be obtained at its arbitrary excitation for an unadjusted slot in the first approximation. The solution of the current integral equation by the iterations method is given in the form of one formula, suitable both for adjusted and unadjusted slots; however, the application of this method is possible only to make concrete the field of the impressed sources at the initial stage of solution. Further, we shall obtain the general analytical expression for the current in the form of one formula, suitable both for resonant and non-resonant slots, without making concrete the field of the impressed sources and the considered electrodynamic volumes by the asymptotic averaging method.

2.3 Asymptotic Averaging Method

The questions of a strict ground of the asymptotic averaging method are a pure mathematial problem investigated in details in the monographs [2.16,2. 17], where the corresponding theorems of averages are proved. Let us consider briefly those whose principals we need further.

Let the system of ordinary differential equations in a standard form (x $-$ the n-dimensional vector, $0 < \alpha \ll 1$ - the small parameter) be set

$$\frac{\mathrm{d}x}{\mathrm{d}s} = \alpha X(s, x), \qquad (2.35)$$

which is characterized by the fact that the first derivatives $\frac{dx}{ds}$ are proportional to the small parameter, that is, x variables change slowly. Different methods exist to reduce the equations to form (2.35), at the same time the method of variation of arbitrary constants is oftener used [2.17]. Considering that reducing of the initial

equations system to the standard form has been done, we change the variables in
(2.35)

$$x = \zeta + \alpha \tilde{X}(s, \zeta),\tag{2.36}$$

where ζ are considered as new unknowns, $\frac{\partial \tilde{X}}{\partial s} = X(s, \zeta) - \overline{X}(\zeta)$, and the bar symbol
means average on the explicitly containing variable s

$$\overline{X}(\zeta) = \lim_{l \to \infty} \frac{1}{l} \int_0^l X(s, \zeta) ds.\tag{2.37}$$

Then, after some transformations, we can get [2.16]

$$\frac{d\zeta}{ds} = \alpha \overline{X}(\zeta) + \alpha^2 ...,\tag{2.38}$$

that is, if ζ satisfies the equations (2.38), the right part of which differs from the
right part of averaging equations

$$\frac{d\zeta}{ds} = \alpha \overline{X}(\zeta)\tag{2.39}$$

at the values of the α^2 order, then the expression (2.36) is an accurate development
of the input equations (2.35). That is why we can accept $x = \zeta$ as the first approximation where ζ is solution of equations (2.39), which satisfies equations (2.35)
precisely to the values of the infinitesimal second order.

Thus, the equations in the first approximation (2.39) are obtained from the exact
equations (2.35) by means of averaging the last on the s-variable and ζ are considered to be constants. This formal process, which consists of changing the exact equations by averaging ones, is called the averaging principle. Its essence was brought to
light by N. Bogolyubov and U. Mitropolsky [2.16], who showed that the averaging
method was connected with the existence of some change of variables which allow
us to omit s from the right part of equations with any accuracy grade relative to
the α-small parameter. It provides the opportunity to build not only the system of
the first approximation (2.39) but to find the averaging system of higher orders,
which, when solved, approximate the solutions of (2.35) with arbitrary fixed preci-
sion though, practically, because of quick complication of formulas the first approximation can effectively be mainly used. The proof of the smallness of the first approximation error was also obtained by N. Bogolyubov and U. Mitropolsky [2.16], who
determined that under rather general conditions the difference $x(s) - \zeta(s)$ could
be made as small as possible for sufficiently small α on any big but finite interval
$0 < s < l$. Thus, the following basic theorem of average takes place [2.16, 2.17].

Theorem 1. Let the function $X(s, x)$ be determined and continuous in domain Q
$(s \geq 0, x \in D)$ and in this domain:

1. $X(s, x) \in \mathrm{Lip}_x(\lambda, Q)$, that is, $X(s, x)$ satisfies the Lipshits condition with the constant λ at x;
2. $||X(s, x)|| < M$, that is, $X(s, x)$ is limited;
3. in each point $x \in D$ the limit (2.37) exists;
4. the solution $\zeta(s)$ of the averaging system (2.39) is determined for all $s \geq 0$ and is situated in the D-domain with some ρ-neighborhood.

Then for any possibly small $\eta > 0$ and any possibly big $L > 0$ we can indicate such α_0, that at $0 < \alpha < \alpha_0$ the unequation will be fulfilled on the segment $0 \leq s \leq L\alpha^{-1}$

$$||x(s) - \zeta(s)|| < \eta,$$

where $x(s)$ and $\zeta(s)$ are the solutions of the systems (2.35) and (2.39), correspondingly, coinciding at $s = 0$.

This theorem is also generalized on the integral–differential equations in a standard form

$$\frac{dx}{ds} = \alpha X \left(s, x, \int_0^s \phi\left(s, s', x(s')\right) ds' \right),$$ (2.40)

for which the following scheme of averages is possible [2.17]. Let us calculate the integral

$$\psi(s, x) = \int_0^s \phi(s, s', x) ds'$$ (2.41)

on the explicitly entering variable s' (here s and x are considered to be parameters). Now alongside with (2.40) let us consider the differential equations system

$$\frac{dy}{ds} = \alpha X\left(s, y, \psi(s, y)\right),$$ (2.42)

which is subject to averaging, namely: let the limit exist

$$\lim_{l \to \infty} \frac{1}{l} \int_0^l X\left(s, y, \psi(s, y)\right) ds = \overline{X}(y).$$ (2.43)

Then we obtain the following differential equations system

$$\frac{d\zeta}{ds} = \alpha \overline{X}(\zeta).$$ (2.44)

The system (2.44) is averaged, corresponding to the integral–differential equations system (2.40). Thus, the idea of average in the systems of the form (2.40) is to approximate solutions of this system by solutions of a specially selected system of differential equations of the form (2.44), which probably is easier to investigate than the initial integral–differential equations system. The conditions under which the proximity of solutions (2.40), (2.42) and (2.44) take place are given in [2.17].

In the integral–differential systems (as distinct from the average in the differential systems), different variants of average are possible. In general, for one system of the integral–differential equations it is possible to put into accordance some different systems of the averaged equations. Some of these averaged systems are the differential equations systems, others—integral–differential equations systems. The possibility of choosing a more suitable averaged system defines high efficiency of the method of average at applied problems solutions.

Let us consider the following integral–differential equations system

$$\frac{dx}{ds} = \alpha X \left(s, x, \frac{dx}{ds}, \int_0^s \phi \left(s, s', x(s'), \frac{dx(s')}{ds'} \right) ds' \right), \qquad (2.45)$$

which is the system of a standard form unsolved relating to the derivative. Practically, the attempt to solve the system relative to $\frac{dx}{ds}$ often necessitates fulfilling intricate and laborious computations. That is the reason for the corresponding schemes of averages, which allow to avoid this [2.17]. Thus, in the first approximation, instead of (2.45) we can consider the simplified system

$$\frac{dy}{ds} = \alpha X \left(s, y, 0, \int_0^s \phi \left(s, s', y(s'), 0 \right) ds' \right), \qquad (2.46)$$

because availability of the derivates in the right part (2.45) begins to affect the second and further asymptotic approximations. The solutions (2.45) and (2.46) are arbitrarily near on the segment of the $L\alpha^{-1}$ order at sufficiently general conditions and small α.

Both in the integral–differential equations systems and in the differential equations ones different variants of partial average are possible, that is, only several addendums or individual equations in the fixed system [2.17] undergo average. Particularly, if the initial system has the form

$$\frac{dx}{ds} = \alpha X_1(s, x) + \alpha X_2(s, x) \qquad (2.47)$$

and the limit exists

$$\lim_{l \to \infty} \frac{1}{l} \int_0^l X_1(s, x) \mathrm{d}s = \overline{X}_1(x), \tag{2.48}$$

then the partially averaged system corresponds to the system (2.47)

$$\frac{\mathrm{d}\zeta}{\mathrm{d}s} = \alpha \overline{X}_1(\zeta) + \alpha X_2(s, \zeta). \tag{2.49}$$

Variants of partial averaging are rather numerous and also stretch on the systems of the kind (2.40) and (2.45).

References

2.1. Bethe, H.A.: Theory of diffraction by small holes. Phys. Rev. **66**, 163–182 (1944)
2.2. Stevenson, A.F.: Theory of slots in rectangular wave-guides. J. Appl. Phys. **19**, 24–38 (1948)
2.3. Lewin, L.: Some observations on waveguide coupling through medium sized slots. Proc. Inst. Elec. Eng. **107C**, 171–178 (1960)
2.4. Sangster, A.J.: Variational method for the analysis of waveguide coupling. Proc. Inst. Elec. Eng. **112**, 2171–2179 (1965)
2.5. Levinson, I.B., Fredberg, P.Sh.: Slot couplers of rectangular one mode waveguide equivalent circuits and lumped parameters. Radio Eng. and Electron. Phys. **11**, 717–724 (1966)
2.6. Khac, T.Vu.: Solutions for some waveguide discontinuities by the method of moments. IEEE Trans. Microwave Theory and Tech. **MTT-20**, 416–418 (1972)
2.7. Rengarajan, S.R.: Analysis of a centered-inclined waveguide slot coupler. IEEE Trans. Microwave Theory and Tech. **MTT-37**, 884–889 (1989)
2.8. Das, B.N., Chakraborty, A., Narasimha Sarma, N.V.S.: S matrix of slot-coupled H-plane Tee junction using rectangular waveguides. IEEE Trans. Microwave Theory and Tech. **MTT-38**, 779–781 (1990)
2.9. Pandharipande, V.M., Das, B.N.: Coupling of waveguides through large apertures. IEEE Trans. Microwave Theory and Tech. **MTT-26**, 209–212 (1978)
2.10. Pandharipande, V.M., Das, B.N.: Equivalent circuit of a narrow-wall waveguide slot coupler. IEEE Trans. Microwave Theory and Tech. **MTT-27**, 800–804 (1979)
2.11. Sangster, A.J., Wang, H.: A hybrid analytical technique for radiating slots in waveguide. Journal of Electromagnetic Waves and Applications. **9**, 735–755 (1995)
2.12. Katrich, V.A., Nesterenko, M.V., Khizhnyak, N.A.: Asymptotic solution of integral equation for magnetic current in slot radiators and coupling apertures. Telecommunications and Radio Engineering. **63**, 89–107 (2005)
2.13. Butler, C.M., Umashankar, K.R.: Electromagnetic excitation of a wire through an aperture-perforated conducting screen. IEEE Trans. Antennas and Propagat. **AP-24**, 456–462 (1976)
2.14. King, R.W.P.: The Theory of Linear Antennas. Harv. Univ. Press, Cambr.-Mass. (1956)
2.15. Naiheng, Y., Harrington, R.: Electromagnetic coupling to an infinite wire through a slot in a conducting plane. IEEE Trans. Antennas and Propagat. **AP-31**, 310–316 (1983)
2.16. Bogolyubov, N.N., Mitropolsky, U.A.: Asymptotic Methods in the Theory Nonlinear Fluctuations. Nauka, Moskow (1974) (in Russian)
2.17. Philatov, A.N.: Asymptotic Methods in the Theory of Differential and Integral-differential Equations. PHAN, Tashkent (1974) (in Russian)

Chapter 3
Analytical Solution of the Integral Equations for the Current by the Averaging Method

In this chapter the asymptotic averaging method has been used to obtain the general approximate analytical expression for the magnetic current in the slot applied both for the adjusted slots ($kL = n\pi/2$, $n = 1, 2, 3, \ldots$) and for the unadjusted ones ($kL \neq n\pi/2$) coupling two waveguides of different cross-section sizes in a common case, which are excited by the arbitrary field of impressed sources.

3.1 Solution of the Equation for the Current in a General Form

Due to the constants variation method [2.16] let us change the variables:

$$J(s) = A(s)\cos ks + B(s)\sin ks, \tag{3.1}$$

$$\frac{dJ(s)}{ds} = -A(s)k\sin ks + B(s)k\cos ks,$$

$$\left(\frac{dA(s)}{ds}\cos ks + \frac{dB(s)}{ds}\sin ks = 0\right),$$

$$\frac{d^2 J(s)}{ds^2} + k^2 J(s) = -\frac{dA(s)}{ds}\sin ks + \frac{dB(s)}{ds}\cos ks$$

$$= \frac{\alpha}{k}\{i\omega H_{0s}(s) + F_N[s, J(s)]\}.$$

Equation (2.8) goes to the next system of integral–differential equations for the unknown functions $A(s)$ and $B(s)$:

$$\frac{dA(s)}{ds} = -\frac{\alpha}{k}\{i\omega H_{0s}(s) + F_N\left[s, A(s), \frac{dA(s)}{ds}, B(s), \frac{dB(s)}{ds}\right]\}\sin ks,$$

$$\frac{dB(s)}{ds} = +\frac{\alpha}{k}\{i\omega H_{0s}(s) + F_N\left[s, A(s), \frac{dA(s)}{ds}, B(s), \frac{dB(s)}{ds}\right]\}\cos ks, \tag{3.2}$$

where $F_N = F + F_0$, and it is the slot full own field.

The equations obtained are fully equivalent to equation (2.8), and they are the system of integral–differential equations of standard type unsolved for a derivative.

M.V. Nesterenko et al., *Analytical and Hybrid Methods in the Theory of Slot-Hole Coupling of Electrodynamic Volumes*, DOI: 10.1007/978-0-387-76362-0_3, © Springer Science+Business Media, LLC, 2008

The right-hand parts of these equations are proportional to the α small parameter. Therefore, the $A(s)$ and $B(s)$ functions in the right-hand parts of the equations (3.2) can be regarded as slowly changing functions. To solve the system of the equations in (3.2) it is possible to use the asymptotic averaging method. Then when we put the system of the equations (3.2) in accordance with the simplified system [2.16], where in the right-hand parts of the equations $\frac{dA(s)}{ds} = 0$, $\frac{dB(s)}{ds} = 0$. When we make partial averaging [2.17] in (3.2) along the s explicitly entering variable we obtain two equations of the first approximation:

$$\frac{d\overline{A}(s)}{ds} = -\alpha \left\{ \frac{i\omega}{k} H_{0s}(s) + \overline{F}_N[s, \overline{A}, \overline{B}] \right\} \sin ks, \qquad (3.3)$$

$$\frac{d\overline{B}(s)}{ds} = +\alpha \left\{ \frac{i\omega}{k} H_{0s}(s) + \overline{F}_N[s, \overline{A}, \overline{B}] \right\} \cos ks,$$

where

$$\overline{F}_N[s, \overline{A}, \overline{B}] = [\overline{A}(s') \sin ks' - \overline{B}(s') \cos ks'] \left. G_s^\Sigma(s, s') \right|_{-L}^{L},$$

$$G_s^\Sigma(s, s') = 4 \frac{e^{-ikR(s,s')}}{R(s, s')} + G_{0s}^e(s, s') + G_{0s}^i(s, s') \qquad (3.4)$$

$$= G_s^e(s, s') + G_s^i(s, s').$$

Integrating the system (3.3) and substituting the obtained values $\overline{A}(s)$ and $\overline{B}(s)$ as the approximating functions for $A(s)$ and $B(s)$ in (3.1) we get the most general asymptotic expression for the narrow slot current in the arbitrary position relative to coupled volumes walls:

$$J(s) = \overline{A}(-L) \cos ks + \overline{B}(-L) \sin ks$$

$$+ \alpha \int_{-L}^{s} \left\{ \frac{i\omega}{k} H_{0s}(s') + \overline{F}_N[s', \overline{A}, \overline{B}] \right\} \sin k(s - s') \, ds'. \qquad (3.5)$$

To define the constants $\overline{A}(\pm L)$ and $\overline{B}(\pm L)$ it is necessary to use the boundary conditions $J(\pm L) = 0$ and the symmetry conditions [2.14], which are uniquely connected with the slot excitation technique. Then taking into consideration the symmetrical (index "s") and the antisymmetrical (index "a") current components at arbitrary excitation $H_{0s}(s) = H_{0s}^s(s) + H_{0s}^a(s)$ of the slot with the accuracy not more than the terms of the α^2 order, we finally have:

$$J(s) = J^s(s) + J^a(s) = \alpha \frac{i\omega}{k}$$

$$\times \left\{ \int_{-L}^{s} H_{0s}(s') \sin k(s - s')ds' - \frac{\sin k(L + s) \int_{-L}^{L} H_{0s}^s(s') \sin k(L - s')ds'}{\sin 2kL + \alpha N^s(kd, 2kL)} \right.$$

$$\left. - \frac{\sin k(L + s) \int_{-L}^{L} H_{0s}^a(s') \sin k(L - s')ds'}{\sin 2kL + \alpha N^a(kd, 2kL)} \right\}.$$

$$(3.6)$$

where $N^s(kd, 2kL)$ and $N^a(kd, 2kL)$ are the functions of the slot own field which are equal, respectively:

$$N^s(kd, 2kL) = \int_{-L}^{L} [G_s^{\Sigma}(s, -L) + G_s^{\Sigma}(s, L)] \sin k(L - s)ds,$$

$$N^a(kd, 2kL) = \int_{-L}^{L} [G_s^{\Sigma}(s, -L) - G_s^{\Sigma}(s, L)] \sin k(L - s)ds, \qquad (3.7)$$

and which are completely defined by Green's functions of the coupled volumes representing infinite and half-infinite waveguides, resonators, etc.

It is necessary to note that near the resonance ($\sin 2kL \approx 0$) the main contribution to the current amplitude is made by the functions of the slot own field $N^s(kd, 2kL)$ and $N^a(kd, 2kL)$, which take into account both the basic oscillation mode and high wave modes in the surroundings of the slot.

3.2 Single Slots in the Common Walls of Rectangular Waveguides

As an example, let us consider the coupling of two identical rectangular waveguides by the $\{a \times b\}$ cross sections through the symmetrical transverse slot in the common broad wall and via the longitudinal slot in the common narrow wall. We also consider the coupling of two mutually perpendicular waveguides in the H-plane through the longitudinal/transverse slot in the common broad wall.

3.2.1 Symmetrical Transverse Slot in a Common Broad Wall of Waveguides

In this case $H_{0s}(s) = H_{0s}^s(s) = H_0 \cos \frac{\pi s}{a}$. Taking into account that for coupling of two waveguides of equal sizes $N^s = 2W^s$, $N^a = 2W^a$, we get:

$$J(s) = -\alpha H_0 \frac{i\omega \left\{\cos ks \cos\left(\frac{\pi L}{a}\right) - \cos kL \cos\left(\frac{\pi s}{a}\right)\right\}}{\gamma^2[\cos kL + \alpha 2 W_t^s(kd, kL)]}, \tag{3.8}$$

where $\gamma^2 = k^2 - (\pi/a)^2$, H_0 is the amplitude of the incident H_{10}-wave, falling from $z = -\infty$ (a region 1).

The $|S_{11}|$ reflection coefficient, the $|S_{12}|$ transmission ones in the first waveguide and the transmission coefficients $|S_{13}|$ and $|S_{14}|$ in the second waveguide are equal, respectively:

$$|S_{11}| = |S_{13}| = |S_{14}| = \left| \frac{4\pi\alpha f\left(kL, \frac{\pi}{a}L\right)}{abk\gamma[\cos kL + \alpha 2 W_t^s(kd, kL)]} \right|,$$

$$|S_{12}| = \left| 1 - \frac{4\pi\alpha f\left(kL, \frac{\pi}{a}L\right)}{iabk\gamma[\cos kL + \alpha 2 W_t^s(kd, kL)]} \right|, \tag{3.9}$$

where

$$f\left(kL, \frac{\pi}{a}L\right) = 2\cos\frac{\pi}{a}L \frac{\sin kL \cos\frac{\pi}{a}L - \frac{\pi}{ka}\cos kL \sin\frac{\pi}{a}L}{1 - (\pi/ka)^2}$$
$$- \frac{\cos kL}{(2\pi/ka)}\left(\sin\frac{2\pi L}{a} + \frac{2\pi L}{a}\right).$$

3.2.2 Longitudinal Slot in a Common Narrow Wall of Waveguides

For a longitudinal slot, the field projection of impressed sources on the slot axis equals $H_{0s}(s) = H_0 \exp(-i\gamma s)$, and we have:

$$J(s) = J^s(s) + J^a(s) = \alpha H_0 \frac{i\omega}{(\pi/a)^2}\left\{ e^{-i\gamma s} \right.$$
$$\left. - \frac{\cos ks \cos \gamma L}{\cos kL + \alpha 2 W_{ln}^s(kd, kL)} + i\frac{\sin ks \sin \gamma L}{\sin kL + \alpha 2 W_{ln}^a(kd, kL)} \right\}. \tag{3.10}$$

For the reflection, transmission and coupling coefficients we obtain the following expressions:

$$|S_{11}| = |S_{13}| = \left| \frac{4\pi\alpha}{abk\gamma} \left\{ \frac{f^s\,(kL,\gamma L)}{\cos kL + \alpha 2 W_{ln}^s\,(kd,kL)} + \frac{f^a\,(kL,\gamma L)}{\sin kL + \alpha 2 W_{ln}^a\,(kd,kL)} - 2kL \frac{\sin 2\gamma L}{2\gamma L} \right\} \right|,$$

$$|S_{12}| = \left| 1 + \frac{i4\pi\alpha}{abk\gamma} \left\{ \frac{f^s\,(kL,\gamma L)}{\cos kL + \alpha 2 W_{ln}^s\,(kd,kL)} + \frac{f^a\,(kL,\gamma L)}{\sin kL + \alpha 2 W_{ln}^a\,(kd,kL)} - 2kL \right\} \right|,$$

$$|S_{14}| = |S_{12} - 1|, \tag{3.11}$$

where

$$f^s\,(kL,\gamma L) = 2\cos\gamma L \frac{\sin kL \cos\gamma L - (\gamma/k)\cos kL \sin\gamma L}{1 - (\gamma/k)^2},$$

$$f^a\,(kL,\gamma L) = 2\sin\gamma L \frac{\cos kL \sin\gamma L - (\gamma/k)\sin kL \cos\gamma L}{1 - (\gamma/k)^2}.$$

3.2.3 Longitudinal/Transverse Slot in a Common Broad Wall of Waveguides

In this case, the current distribution depends upon where the waveguide excitation sources are situated. If the H_{10} incident wave is propagating in the waveguide for which the coupling slot is transverse, then we get:

$$J(s) = -\alpha H_0 \frac{i\omega \left\{ \cos ks \cos\left(\frac{\pi L}{a}\right) - \cos kL \cos\left(\frac{\pi s}{a}\right) \right\}}{\gamma^2 \left\{ \cos kL + \alpha [W_t^s\,(kd,kL) + W_{lb}^s(kd,kL)] \right\}}. \tag{3.12}$$

If the slot for the exciting field is longitudinal, then we have:

$$J(s) = J^s(s) + J^a(s)$$

$$= -\alpha H_0 \frac{i\omega \cos\dfrac{\pi x_0}{a}}{(\pi/a)^2} \left\{ \frac{(\cos ks \cos\gamma L - \cos kL \cos\gamma s)}{\cos kL + \alpha [W_{lb}^s(kd,kL) + W_t^s(kd,kL)]} \right.$$
$$\left. - i\frac{(\sin ks \sin\gamma L - \sin kL \sin\gamma s)}{\sin kL + \alpha [W_{lb}^a(kd,kL) + W_t^a(kd,kL)]} \right\}. \tag{3.13}$$

For the current in (3.12) the coupling coefficients are defined by the expressions (3.9), where it is necessary to make the following change $2W_t^s \to W_t^s + W_{lb}^s$. For the current in (3.13) they equal, respectively, (the H_{10} incidence wave spreads from the region 3 into the region 4):

$$|S_{33}| = \left| \frac{4\pi\alpha \cos^2 \frac{\pi x_0}{a}}{abk\gamma} \left\{ \frac{f_1^s (kL, \gamma L)}{\cos kL + \alpha[W_{lb}^s (kd, kL) + W_t^s (kd, kL)]} + \frac{f_1^a (kL, \gamma L)}{\sin kL + \alpha[W_{lb}^a (kd, kL) + W_t^a (kd, kL)]} \right\} \right|,$$

$$|S_{34}| = \left| 1 - \frac{4\pi\alpha \cos^2 \frac{\pi x_0}{a}}{iabk\gamma} \left\{ \frac{f_1^s (kL, \gamma L)}{\cos kL + \alpha[W_{lb}^s (kd, kL) + W_t^s (kd, kL)]} - \frac{f_1^a (kL, \gamma L)}{\sin kL + \alpha[W_{lb}^a (kd, kL) + W_t^a (kd, kL)]} \right\} \right|,$$

$$|S_{31}|^2 = |S_{32}|^2 = \frac{1}{2}(1 - |S_{33}|^2 - |S_{34}|^2), \tag{3.14}$$

where

$$f_1^s (kL, \gamma L) = f^s (kL, \gamma L) - \frac{\cos kL}{2 (\gamma/k)} (\sin 2\gamma L + 2\gamma L),$$

$$f_1^a (kL, \gamma L) = f^a (kL, \gamma L) - \frac{\sin kL}{2 (\gamma/k)} (\sin 2\gamma L - 2\gamma L).$$

The expressions for $W_t^s (kd, kL)$, $W_t^a (kd, kL)$, $W_{ln}^s (kd, kL)$, $W_{ln}^a (kd, kL)$, $W_{lb}^s (kd, kL)$, $W_{lb}^a (kd, kL)$ functions are represented in Appendix B.

3.3 Finite Thickness of the Coupling Region Account

The h finite thickness of the wall between coupled volumes can be taken into account due to [3.1, 3.2] which introduces the d_e slot effective width concept. Then at the $(h/\lambda) \ll 1$, precisely up to the terms of the $\{(hd)/\lambda^2\}$ order, we have [3.1]:

$$\frac{h}{d} \ll 1 : d_e = d \left(1 - \frac{1}{\pi} \frac{h}{d} \ln \frac{d}{h} \right);$$

$$\frac{h}{d} \gtrsim 1 : d_e = d \left(\frac{8}{\pi e} e^{-\frac{\pi}{2} \frac{h}{d}} \right). \tag{3.15}$$

The expression in [3.2] is in good approximation for both cases:

$$d_e \cong d\, e^{-\frac{\pi h}{2d}}. \tag{3.16}$$

The $|S_{13}|$ coefficient calculations using the approximate ratios of (3.15) and (3.16) within the limits of $0 \leq {}^h/_{2L} \leq 0.2$ coincide with the result, obtained in [3.3], where the account of the rectangular waveguide wall thickness has been made by solving two coupling integral equations with moments method.

3.4 Numerical Results

The plots of the dependences of the $|S_\Sigma|^2 = |S_{13}|^2 + |S_{14}|^2$ coupling coefficient from the length of the symmetrical transverse slot in the common infinitely thin broad wall of two identical rectangular waveguides are in Fig. 3.1. The dependencies have been calculated by different methods. It is seen that the calculations made by means of the averaging, variational [2.4] and moments methods [2.6] give the values of the slot resonance length (slot "shortening") of $2L \cong 0.47\lambda$. The results obtained by using the quasi-static [2.3] and equivalent circuits (the "reaction" method [2.9]) methods lead to the resonance value of $2L = 0.5\lambda$, which does not correspond to reality. We note that Fig. 3.1 gives the results of solving the problem with the help of the moments method as a numerical experiment.

In Fig. 3.2 are the plots of the amplitude-phase distribution $J(s) = |J(s)| e^{i \arg J(s)}$ of the current along the longitudinal/transverse slot in the common infinitely thin broad wall of two mutually perpendicular waveguides.

It can be seen that if the slot for the H_{10} excitation wave is transverse ($W_t \rightarrow W_{lb}$), then the amplitude distribution of current is purely symmetrical, and the current phase is constant along the slot length. For another case of the excitation ($W_{lb} \rightarrow W_t$), the amplitude distribution of the current is sufficiently asymmetric, and the current phase changes along the slot. The current distribution curves are

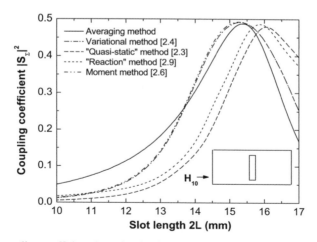

Fig. 3.1 The coupling coefficient dependencies from the symmetrical transverse slot length in the common broad wall of two rectangular waveguides at: $a=22.86$ mm, $b=10.16$ mm, $d=1.5875$ mm, $\lambda=32.0$ mm, $h=0.0$ mm

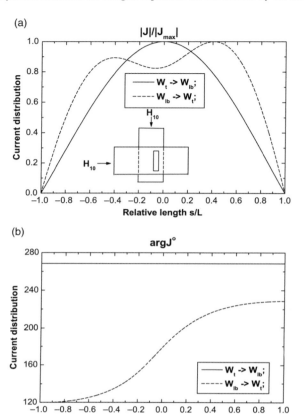

Fig. 3.2 The current distribution along the longitudinal/transverse (W_{lb} and W_t respectively) slots in the common broad wall of two mutually perpendicular waveguides at: $a = 22.86$ mm, $b = 10.16$ mm, $d = 1.5875$ mm, $\lambda = 25.8$ mm, $2L = 20.0$ mm, $x_0 = 1.43$ mm is the slot axis position, $h = 0.0$ mm

of the same kind in the case of two waveguides coupling through the longitudinal slot in the common broad wall of the finite thickness of the rectangular waveguides in Fig. 3.3. The calculations have been made according to the formula (3.13) (where there is the $W_t^{s,a} \rightarrow W_{lb}^{s,a}$ change) and by Galerkin's method:

$$J(s) = \sum_{n=1}^{N} J_n \sin \frac{n\pi(L+s)}{2L},$$ taking into account 2, 6 and 12 basis functions. We can see that the change of the external volume sufficiently increases the contribution of the $J^a(s)$ current antisymmetrical component into the general distribution function.

Thus, the obtained asymptotic solution of the integral equation concerning the magnetic current in slot-hole coupling apertures allows us to obtain analytical expressions for the current to the first approximation, which is valid for various ratios between the wavelength and a longitudinal size of the slot. The given numerical results demonstrate the efficiency and the effectiveness of such solu-

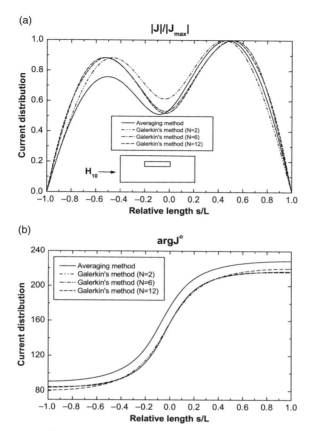

Fig. 3.3 The current distribution along the longitudinal slot in the common broad wall of two rectangular waveguides at: $a = 22.86$ mm, $b = 10.16$ mm, $d = 1.5875$ mm, $\lambda = 25.8$ mm, $2L = 20.0$ mm, $x_0 = 1.43$ mm is the slot axis position, $h = 2.0$ mm

tion. However, there are some quantitative differences between the calculated values of electrodynamics characteristics of the slot coupling apertures, which have been obtained by using the above-mentioned asymptotic formulas and numerical methods. These differences may be removed by using the magnetic current expression obtained from asymptotic solution of the integral equation with the help of the averaging method in combination with other analytical methods, for example, with the induced magnetomotive forces method as was proposed by the authors in [3.4, 3.5].

References

3.1. Garb, H.L., Levinson, I.B., Fredberg, P.Sh.: Effect of wall thickness in slot problems of electrodynamics. Radio Eng. Electron Phys. **13**, 1888–1896 (1968)
3.2. Warne, L.K.: Eddy current power dissipation at sharp corners: closely spaced rectangular conductors. Journal of Electromagnetic Waves and Applications. **9**, 1441–1458 (1995)

3.3. Khac, T.Vu., Carson, C.T.: Coupling by slots in rectangular waveguides with arbitrary wall thickness. Electronics Letters. **8**, 456–458 (1972)

3.4. Berdnik, S.L., Katrich, V.A., Kiiko, V.I., Nesterenko, M.V.: Electrodynamic characteristics of the nonresonant system of transverse slots in the wide wall of rectangular waveguide. Radioelectronics and Communications Systems. **48**, 52–57 (2005)

3.5. Nesterenko, M.V., Katrich, V.A.: The asymptotic solution of an integral equation for magnetic current in a problem of waveguides coupling through narrow slots. Progress In Electromagnetics Research. **PIER 57**, 101–129 (2006)

Chapter 4
Induced Magnetomotive Forces Method for Analysis of Coupling Slots in Waveguides

While applying the moments and Galerkin's methods (see Sect. 1.4) to analyze single and multi-slot elements of the coupling of electrodynamic volumes, different basis and weight functions can be used: the piecewise [2.6], the piecewise linear and piecewise sinusoidal [4.1], the trigonometric [2.7, 2.8, 4.2, 4.3] and Gegenbauer polynomials [4.4]. In these cases it is necessary to solve the system of linear algebraic equations (SLAE) of the N-order, where N is a number of linearly independent basis functions. The system matrix elements (N^2-in all) cannot always be obtained analytically, and calculation time increases proportionally to N^3 [1.4].

For a slots system, the order of SLAE increases proportionally to the number of slots. Therefore it is necessary, to our minds, to approximate the distribution of the slot equivalent magnetic current by one or two functions (it depends on an excitation character) as it was done, for example, in [2.5, 2.9, 2.10]. When only one approximating function exists for every slot in a multi-slot system, the Galerkin's method gets the name the "induced magnetomotive forces method" (IMMFM). In this case the solution is more accurate when the approximating functions describing the slot magnetic current distribution are more accurate. In [2.5, 2.9, 2.10] the half-wave and wave sinusoidal functions were used for the slot with the $2L \leq \lambda$ length. For longer slots it is necessary to increase the number of approximating functions.

We suggest better functions of the current distribution for IMMFM that give sufficiently satisfactory approximation of the current in longer longitudinal slots and in the transverse and longitudinal slots system. These functions have been obtained in Chap. 3 (formula (3.8) for transverse slots and formula (3.13) for longitudinal slots) when we solved the integral equation for the slot magnetic current by the asymptotic averaging method.

4.1 Electrically Long Longitudinal Slot in a Common Broad Wall of Waveguides

Generally, the projection of the $H_{0s}(s)$ impressed field on the slot axis and the $J(s)$ magnetic current in it can be represented with two components—symmetrical and antisymmetrical ones along the slot with the respect to its center—$H_{0s}(s) =$

M.V. Nesterenko et al., *Analytical and Hybrid Methods in the Theory of Slot-Hole Coupling of Electrodynamic Volumes,* DOI: 10.1007/978-0-387-76362-0_4,
© Springer Science+Business Media, LLC, 2008

$H_{0s}^s(s) + H_{0s}^a(s)$, $J(s) = J^s(s) + J^a(s)$. Owing to this, we can have the following integral–differential equation for the current (2.5) in a narrow linear slot:

$$
\left(\frac{d^2}{ds^2} + k^2\right) \int_{-L}^{L} [J^s(s') + J^a(s')] [G_s^e(s, s') + G_s^i(s, s')] \, ds'
$$
$$
= -i\omega [H_{0s}^s(s) + H_{0s}^a(s)].
$$
(4.1)

Let us represent the current as unknown amplitudes and distribution functions fixed:

$$
J(s) = J_0^s f^s(s) + J_0^a f^a(s),
$$
(4.2)

where the $f^s(s)$ and $f^a(s)$ functions must satisfy the following boundary conditions: $f^s(\pm L) = 0$, $f^a(\pm L) = 0$. From (4.2) we have only two unknown amplitudes J_0^s and J_0^a. They can be obtained from two independent equations with respect to J_0^s and J_0^a that we have, using IMMFM:

$$
J_0^s[Y_s^e(kd, kL) + Y_s^i(kd, kL)] = -\frac{i\omega}{2k} M_s(kL),
$$
$$
J_0^a[Y_a^e(kd, kL) + Y_a^i(kd, kL)] = -\frac{i\omega}{2k} M_a(kL),
$$
(4.3)

where

$$
Y_{s,a}^{e,i}(kd, kL) = \frac{1}{2k} \int_{-L}^{L} f^{s,a}(s) \left[\left(\frac{d^2}{ds^2} + k^2\right) \int_{-L}^{L} f^{s,a}(s') G_{s,a}^{e,i}(s, s') \, ds'\right] ds
$$
(4.4)

are the external and inner partial slot admittances and

$$
M_{s,a}(kd, kL) = \int_{-L}^{L} f^{s,a}(s) H_{0s}^{s,a}(s) \, ds
$$
(4.5)

are the partial magnetomotive forces.

For the longitudinal slot in the broad wall of the rectangular waveguide due to (3.13), the $f^s(s)$ and $f^a(s)$ basis functions have the following forms:

$$
f^s(s) = \cos ks \cos \gamma L - \cos kL \cos \gamma s,
$$
$$
f^a(s) = \sin ks \sin \gamma L - \sin kL \sin \gamma s.
$$
(4.6)

Using (4.6) we can obtain unknown amplitudes J_0^s, J_0^a from (4.3). It gives us the opportunity to obtain energy characteristics of the coupling slot elements.

For the longitudinal slot in the common broad wall of rectangular waveguides, we have:

$$|S_{11}| = |S_{13}| = \left| \frac{2\pi^3 \cos^2 \frac{\pi x_0}{a}}{a^3 b \gamma k^3} \left[\tilde{J}_0^s F^s(kL) + \tilde{J}_0^a F^a(kL) \right] \right| ,$$

$$|S_{12}| = \left| 1 - \frac{2\pi^3 \cos^2 \frac{\pi x_0}{a}}{i a^3 b \gamma k^3} \left[\tilde{J}_0^s F^s(kL) - \tilde{J}_0^a F^a(kL) \right] \right| ,$$

$$|S_{14}| = |S_{12} - 1| , \tag{4.7}$$

where

$$\tilde{J}_0^s = \frac{F^s(kL)}{Y_s^e(kd, kL) + Y_s^i(kd, kL)}, \qquad \tilde{J}_0^a = -\frac{F^a(kL)}{Y_a^e(kd, kL) + Y_a^i(kd, kL)},$$

$$F^s(kL) = 2 \cos \gamma L \frac{\sin kL \cos \gamma L - (\gamma/k) \cos kL \sin \gamma L}{(\pi/ka)^2} - \cos kL \frac{\sin 2\gamma L + 2\gamma L}{2(\gamma/k)},$$

$$F^a(kL) = 2 \sin \gamma L \frac{\cos kL \sin \gamma L - (\gamma/k) \sin kL \cos \gamma L}{(\pi/ka)^2} - \sin kL \frac{\sin 2\gamma L - 2\gamma L}{2(\gamma/k)}.$$

$$\tag{4.8}$$

The expressions for the $Y_s^{i,e}(kd, kL)$, $Y_a^{i,e}(kd, kL)$ admittances are given in Appendix C.

Generally, the slot can be located at the φ angle to the longitudinal waveguide axis. Then according to the general solution of the integral equation (3.6) for the current, the basis functions of IMMFM have the forms:

$$f^s(s) = \frac{\cos ks \cos k_2 L - \cos kL \cos k_2 s}{(\sin \varphi + (k_c/\gamma) \cos \varphi)^2} e^{ik_c x_0}$$
$$- \frac{\cos ks \cos k_1 L - \cos kL \cos k_1 s}{(\sin \varphi - (k_c/\gamma) \cos \varphi)^2} e^{-ik_c x_0},$$

$$f^a(s) = \frac{\sin ks \sin k_2 L - \sin kL \sin k_2 s}{(\sin \varphi + (k_c/\gamma) \cos \varphi)^2} e^{ik_c x_0}$$
$$+ \frac{\sin ks \sin k_1 L - \sin kL \sin k_1 s}{(\sin \varphi - (k_c/\gamma) \cos \varphi)^2} e^{-ik_c x_0}, \tag{4.9}$$

where: $k_1 = k_c \sin \varphi + \gamma \cos \varphi$, $k_2 = k_c \sin \varphi - \gamma \cos \varphi$, x_0 is the distance between the narrow waveguide wall and the slot center. Let us note that at $\varphi = 0$ formulas (4.9) are transformed into (4.6).

The current distribution curves in the case of two rectangular waveguides coupled through the longitudinal slot in the common broad wall of the finite thickness are given in Fig. 4.1. The calculations have been made according to formulas (4.2) and (4.6) and by Galerkin's method, taking into account 6 basis functions. In Fig. 4.2, the plots of the $|S_\Sigma|^2 = |S_{13}|^2 + |S_{14}|^2$ coupling coefficients dependences of the longitudinal slot in the broad waveguide wall due to its electrical length are represented.

Fig. 4.1 The current distribution along the longitudinal slot in the common broad wall of two rectangular waveguides at: $a = 22.86$ mm, $b = 10.16$ mm, $d = 1.5875$ mm, $\lambda = 25.8$ mm, $2L = 20.0$ mm, $x_0 = 1.43$ mm is the slot axis position, $h = 2.0$ mm

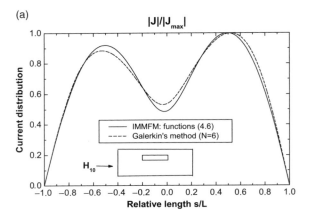

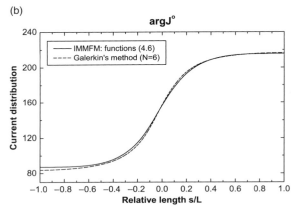

The calculations have been made with the following representations of the magnetic current: a) in the form of $J(s) = \sum_{n=1}^{N} J_n \sin \frac{n\pi(L+s)}{2L}$ (Galerkin's method); b) with the use of the functions (4.6), c) using the following approximation [2.5]:

$$J(s) = J_0^s \cos\left(\frac{\pi s}{2L}\right) + J_0^a \sin\left(\frac{\pi s}{L}\right). \tag{4.10}$$

Figure 4.2 also gives the calculated values obtained by using the finite elements method (FEM) due to the program "CST Microwave Studio".

Hence, the approximation (4.6), when only two basis functions are used, gives good match with the results obtained by using Galerkin's and the finite elements methods for the longitudinal slots with the electrical length up to $2L/\lambda \leq 2.75$ (when it is necessary to use 12 basis functions for Galerkin's method); meanwhile the (4.10) approximation is satisfactory for only up to $2L/\lambda \leq 1.25$. We think that (4.10) functions describe the slot current distribution less accurately

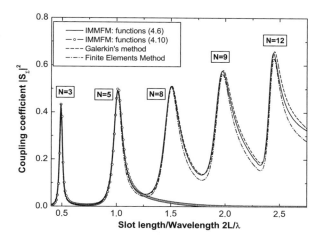

Fig. 4.2 The coupling coefficient dependences from the relative length of the longitudinal slot in the common broad wall of two rectangular waveguides at: $a = 23.0$ mm, $b = 10.0$ mm, $d = a/15$, $x_0 = a/6$, $\lambda/\lambda_c = 0.625$, $h = 2.0$ mm

than the (4.6) functions because they do not have the λ, λ_c (cut-off H_{10} wavelength) and λ_g (wavelength in the waveguide) values in the distribution, the ratios of which between each other and slot length suppose the formation of the magnetic current amplitude–phase distribution and energy characteristics of the coupling slot-holes.

The latter is proved by the plots given in Fig. 4.3 where the curves of the coupling coefficient dependences of two identical waveguides through the longitudinal slot in their common broad wall from $2L/\lambda$ are represented at different values of λ/λ_c. It is clear if the λ/λ_c ratio increases, then the Q-factor of the $|S_\Sigma|^2 = f(2L/\lambda)$ resonance curves decreases. At a definite ratio between $2L/\lambda$ and λ/λ_c, practically, the full power of the initial wave from one waveguide to the other one can be transmitted.

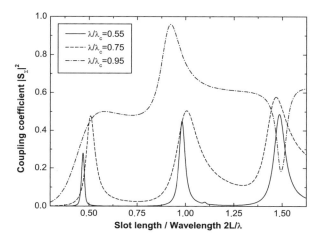

Fig. 4.3 The coupling coefficient dependences from the relative length of the longitudinal slot in the common broad wall of two rectangular waveguides at: $a = 23.0$ mm, $b = 10.0$ mm, $d = a/15$, $x_0 = a/4$, $h = 2.0$ mm

4.2 Two Symmetrical Transverse Slots in a Common Broad Wall of Waveguides

For the two slots in a waveguide wall (Fig. 4.4), one can obtain the system of two coupled integral–differential equations for the $J_1(s_1)$ and $J_2(s_2)$ magnetic currents in the first and the second slots:

$$
\begin{cases}
\left(\dfrac{d^2}{ds_1^2}+k^2\right)\left[\displaystyle\int_{-L_1}^{L_1} J_1(s_1')G_{s_1}^{\Sigma}(s_1,s_1')ds_1' + \displaystyle\int_{-L_2}^{L_2} J_2(s_2')G_{s_1}^{\Sigma}(s_1,s_2')ds_2'\right] = -i\omega H_{0s_1}(s_1), \\[4mm]
\left(\dfrac{d^2}{ds_2^2}+k^2\right)\left[\displaystyle\int_{-L_2}^{L_2} J_2(s_2')G_{s_2}^{\Sigma}(s_2,s_2')ds_2' + \displaystyle\int_{-L_1}^{L_1} J_1(s_1')G_{s_2}^{\Sigma}(s_2,s_1')ds_1'\right] = -i\omega H_{0s_2}(s_2),
\end{cases}
$$
$$(4.11)$$

where $G_{s_{1,2}}^{\Sigma}(s_{1,2},s_{1,2}') = G_{s_{1,2}}^{e}(s_{1,2},s_{1,2}') + G_{s_{1,2}}^{i}(s_{1,2},s_{1,2}')$, $H_{0s_1}(s_1)$ and $H_{0s_2}(s_2)$ are the projections of the field of the impressed sources to the slot axes.

As in a previous case (in Sect. 4.1) the currents in every slot can be written in the following way:

$$J_1(s_1) = J_{01}\,f_1(s_1),\quad J_2(s_2) = J_{02}\,f_2(s_2);\quad f_1(\pm L_1)=0,\quad f_2(\pm L_2)=0.$$
$$(4.12)$$

Due to the induced magnetomotive forces method used for the two slots system, we transform (4.11) into the following algebraic equations system relative to J_{01} and J_{02} unknown amplitudes:

$$
\begin{cases}
J_{01}Y_{11}^{\Sigma}(kL_1,kL_1) + J_{02}Y_{12}^{\Sigma}(kL_1,kL_2) = -\dfrac{i\omega}{2k}M_1(kL_1), \\[4mm]
J_{01}Y_{21}^{\Sigma}(kL_2,kL_1) + J_{02}Y_{22}^{\Sigma}(kL_2,kL_2) = -\dfrac{i\omega}{2k}M_2(kL_2).
\end{cases}
$$
$$(4.13)$$

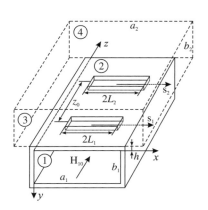

Fig. 4.4 The two symmetrical transverse slots system in the waveguides common wall

Here

$$Y^{\Sigma}_{mn}(kL_m, kL_n) = \frac{1}{2k} \int\limits_{-L_m}^{L_m} f_m(s_m) \left[\left(\frac{d^2}{ds_m^2} + k^2 \right) \int\limits_{-L_n}^{L_n} f_n(s'_n) G^{\Sigma}_{s_m}(s_m, s'_n) ds'_n \right] ds_m,$$

$$(m, n = 1, 2) \tag{4.14}$$

are the eigen $(m = n)$ and mutual $(m \neq n)$ slots admittances, respectively;

$$M_m(kL_m) = \int\limits_{-L_m}^{L_m} f_m(s_m) H_{0sm}(s_m) ds_m, \quad (m = 1, 2) \tag{4.15}$$

are the magnetomotive forces.

In the case of two symmetrical transverse slots, the currents in them are also symmetrical. They can be represented as ($k_c = 2\pi/\lambda_c$, λ_c is the cut-off H_{10} wavelength) due to (3.8):

$$f_m(s_m) = \cos k s_m \cos k_c L_m - \cos k L_m \cos k_c s_m, \quad (m = 1, 2) \tag{4.16}$$

Using (4.16) we can obtain unknown amplitudes J_{01}, J_{02} from (4.13). It gives us the opportunity to obtain energy characteristics of the coupling slots elements:

$$|S_{11}| = |S_{13}| = \left| \frac{2\pi \gamma}{abk^3} \left[\tilde{J}_1 F(kL_1) + e^{-i\gamma z_0} \tilde{J}_2 F(kL_2) \right] \right|,$$

$$|S_{12}| = \left| 1 - \frac{2\pi \gamma}{iabk^3} \left[\tilde{J}_1 F(kL_1) + e^{i\gamma z_0} \tilde{J}_2 F(kL_2) \right] \right|,$$

$$|S_{14}| = |S_{12} - 1|, \tag{4.17}$$

where

$$\tilde{J}_1 = \frac{F(kL_1) Y^{\Sigma}_{22}(kd_2, kL_2) - e^{-i\gamma z_0} F(kL_2) Y^{\Sigma}_{12}(kL_1, kL_2)}{Y^{\Sigma}_{11}(kd_1, kL_1) Y^{\Sigma}_{22}(kd_2, kL_2) - [Y^{\Sigma}_{12}(kL_1, kL_2)]^2},$$

$$\tilde{J}_2 = \frac{e^{-i\gamma z_0} F(kL_2) Y^{\Sigma}_{11}(kd_1, kL_1) - F(kL_1) Y^{\Sigma}_{12}(kL_1, kL_2)}{Y^{\Sigma}_{11}(kd_1, kL_1) Y^{\Sigma}_{22}(kd_2, kL_2) - [Y^{\Sigma}_{12}(kL_1, kL_2)]^2}, \tag{4.18}$$

$$F(kL_m) = 2 \cos k_c L_m \frac{\sin k L_m \cos k_c L_m - (k_c/k) \cos k L_m \sin k_c L_m}{1 - (k_c/k)^2}$$

$$- \cos k L_m \frac{\sin 2k_c L_m + 2k_c L_m}{(2k_c/k)}.$$

Fig. 4.5 The coupling coefficient dependences from the wavelength for a pair of transverse slots in the common broad wall between a pair of rectangular waveguides at: $a = 23.0$ mm, $b = 10.0$ mm, $d_1 = d_2 = 2.0$ mm, $2L_1 = 2L_2 = 10.6$ mm, $z_0 = 10.0$ mm, $h = 1.0$ mm

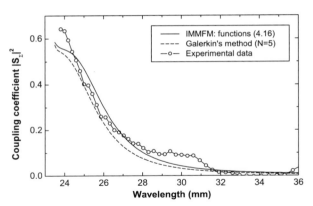

The expressions for the $Y_{mm}(kd_m, kL_m)$, $Y_{mn}(kL_m, kL_n)$ admittances are given in Appendix C.

In Figs. 4.5 and 4.6 the coupling coefficients dependences are given for the system of two identical, rectangular symmetrical transverse slots where the distance between them equals z_0. The calculations have been made with the following representations of the magnetic current: a) in the form of $J_m(s) = \sum_{n=1}^{N} J_{mn} \sin \frac{n\pi(L+s)}{2L}$ (Galerkin's method); b) with the use of the functions (4.16), c) using the following approximation:

$$J_m(s) = J_{0m} \cos\left(\frac{\pi s_m}{2L_m}\right), \quad m = 1, 2. \tag{4.19}$$

It is seen that in the case of two transverse slots, the (4.16) approximation is good. It gives satisfactory coincidence with the results of Galerkin's method and the experimental data in different parts of the band of the operating length of the H_{10}-wave, especially at the resonance points. The (4.19) basis functions change the resonance frequency values.

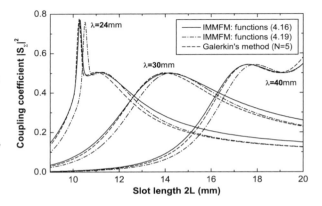

Fig. 4.6 The coupling coefficient dependences from the slot length for a pair of transverse slots in the common broad wall between a pair of rectangular waveguides at: $a = 23.0$ mm, $b = 10.0$ mm, $d_1 = d_2 = 1.6$ mm, $2L_1 = 2L_2 = 2L$, $z_0 = 2\lambda/3$, $h = 0.0$ mm

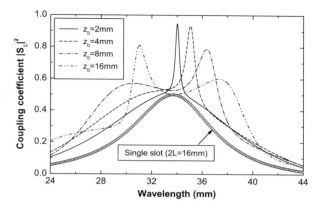

Fig. 4.7 The coupling coefficient dependences from the wavelength for a pair of transverse slots in the common broad wall between a pair of rectangular waveguides at: $a = 23.0$ mm, $b = 10.0$ mm, $d_1 = d_2 = 1.6$ mm, $2L_1 = 2L_2 = 16.0$ mm, $h = 0.0$ mm

The energy characteristics of the slots system are represented in Figs. 4.7–4.10 in dependence from their mutual locations and the wavelength. To be more certain let us consider that $\lambda_{av}=32.53$mm is the average wavelength of a single-mode range of the waveguide at $\lambda_g = \lambda_c$.

The analysis of the plots and expressions (4.17) and (4.18) for the currents and energy characteristics of slots system allows us to make the following conclusions.

- A sufficient increase of the $|S_\Sigma|^2$ coupling coefficient in the narrow range of frequencies (Fig. 4.7) is observed at the distances $z_0 < (\lambda_{av}/2)$ because of strong coupling between the slots along the higher oscillation modes. We note that if $z_0 << (\lambda_{av}/2)$, then $\mathrm{Re}\tilde{J}_1 = -\mathrm{Re}\tilde{J}_2$, $\mathrm{Im}\tilde{J}_1 = \mathrm{Im}\tilde{J}_2$ (moreover $\mathrm{Im}\tilde{J} << \mathrm{Re}\tilde{J}$) and due to (4.17) $|S_{11}| = |S_{12}| = |S_{13}| \cong 0$, $|S_{14}| \cong 1$, that is, practically, the full-power of the incident wave enters to the region 4 ($z_0 = 2$ mm in Fig. 4.7).
- If the distance between the slots $z_0 \geq (\lambda_{av}/2)$ and $2L_1 = 2L_2$, then $|S_\Sigma|^2$ does not depend on z_0 at the resonant wavelength λ_{res} of a single slot (Fig. 4.8, the curve for $\lambda = 33.7$ mm).

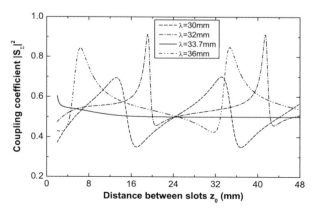

Fig. 4.8 The coupling coefficient dependences from the distance between the slots for a pair of transverse slots in the common broad wall between a pair of rectangular waveguides at: $a = 23.0$ mm, $b = 10.0$ mm, $d_1 = d_2 = 1.6$ mm, $2L_1 = 2L_2 = 16.0$ mm, $h = 0.0$ mm

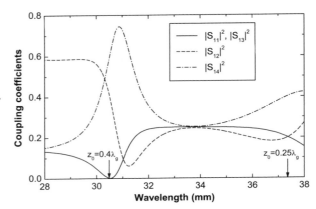

Fig. 4.9 The coupling coefficients dependences from the wavelength for a pair of transverse slots in the common broad wall between a pair of rectangular waveguides at: $a = 23.0$ mm, $b = 10.0$ mm, $d_1 = d_2 = 1.6$ mm, $2L_1 = 2L_2 = 16.0$ mm, $z_0 = 16.0$ mm, $h = 0.0$ mm

- At $z_0 = (n\lambda_g^{res}/2)$ ($n = 1, 2, 3...$, λ_g^{res} is the wavelength in the waveguide, corresponding to the resonant wavelength of a single slot) it follows from (4.17) that in the case of $2L_1 = 2L_2$ the coupling coefficient $|S_\Sigma|^2 = 0.5$, and it does not depend on the wavelength in the neighborhood of the resonance of a single slot (Fig. 4.8).
- Because of strong mutual influence of the slots on each other the minimal value of the reflection coefficient $|S_{11}|^2$ is reached at $z_0 = 0.4\lambda_g$ and not for $z_0 = 0.25\lambda_g$, as it was when the interaction was absent (Fig. 4.9). In this case $|S_{12}|^2 = |S_{14}|^2 = 0.5$, that is, full incident power is divided equally between regions 2 and 4.
- If $z_0 = \lambda_g^{res}/2 = 24.8$ mm ($\lambda_{res} = 33.7$ mm for $2L = 2L_1 = 2L_2 = 16$ mm), then power entering region 1 of the main waveguide, is divided into four equal parts in the wide range of the wavelengths: $|S_{11}|^2 = |S_{12}|^2 = |S_{13}|^2 = |S_{14}|^2 = 0.25$ (Fig. 4.10).

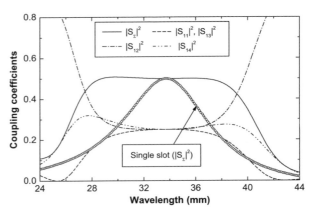

Fig. 4.10 The coupling coefficients dependences from the wavelength for a pair of transverse slots in the common broad wall between a pair of rectangular waveguides at: $a = 23.0$ mm, $b = 10.0$ mm, $d_1 = d_2 = 1.6$ mm, $2L_1 = 2L_2 = 16.0$ mm, $z_0 = 24.8$ mm, $h = 0.0$ mm

4.3 Two Longitudinal Slots in a Common Broad Wall of Waveguides

For the two longitudinal slots in the broad wall of a waveguide (Fig. 4.11) the system of coupled integral–differential equations has the following form concerning the magnetic currents $J_1(s_1)$ and $J_2(s_2)$:

$$
\begin{cases}
\left(\dfrac{d^2}{ds_1^2} + k^2\right)\left[\displaystyle\int_{-L_1}^{L_1} J_1(s_1')G_{s_1}^{\Sigma}(s_1,s_1')ds_1' + \int_{-L_2}^{L_2} J_2(s_2')G_{s_1}^{\Sigma}(s_1,s_2')ds_2'\right] \\
\qquad\qquad\qquad = -i\omega[H_{0s_1}^s(s_1) + H_{0s_1}^a(s_1)], \\[2mm]
\left(\dfrac{d^2}{ds_2^2} + k^2\right)\left[\displaystyle\int_{-L_2}^{L_2} J_2(s_2')G_{s_2}^{\Sigma}(s_2,s_2')ds_2' + \int_{-L_1}^{L_1} J_1(s_1')G_{s_2}^{\Sigma}(s_2,s_1')ds_1'\right] \\
\qquad\qquad\qquad = -i\omega[H_{0s_2}^s(s_2) + H_{0s_2}^a(s_2)].
\end{cases}
\tag{4.20}
$$

Here $G_{s_m}^{\Sigma}(s_m,s_n') = G_{s_m}^{i}(s_m,s_n') + G_{s_m}^{e}(s_m,s_n')$ $(m,n = 1,2)$, $H_{0s_m}^{s,a}$ are the projections of symmetrical (index "s") and antisymmetrical (index "a") components of the magnetic field of impressed sources to the slot axes.

As in the case of a single longitudinal slot (see Sect. 4.1) slot currents can be represented in the form of unknown amplitudes and fixed distribution functions:

$$
J_1(s_1) = J_{01}^s f_1^s(s_1) + J_{01}^a f_1^a(s_1), \quad J_2(s_2) = J_{02}^s f_2^s(s_2) + J_{02}^a f_2^a(s_2),
\tag{4.21}
$$

and the functions $f_m^{s,a}(s_m)$ must also satisfy the edge conditions $f_m^{s,a}(\pm L_m) = 0$.

Substituting (4.21) in the equations (4.20) we obtain the system of linear algebraic equations concerning the unknown amplitudes of the currents $J_{0m}^{s,a}$:

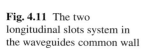

Fig. 4.11 The two longitudinal slots system in the waveguides common wall

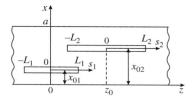

$$\begin{cases} J_{01}^s Y_{11}^{\Sigma ss}(kL_1, kL_1) + J_{02}^s Y_{12}^{\Sigma ss}(kL_1, kL_2) + J_{02}^a Y_{12}^{\Sigma sa}(kL_1, kL_2) = -\dfrac{i\omega}{2k} M_1^s(kL_1), \\[2mm] J_{01}^a Y_{11}^{\Sigma aa}(kL_1, kL_1) + J_{02}^s Y_{12}^{\Sigma as}(kL_1, kL_2) + J_{02}^a Y_{12}^{\Sigma aa}(kL_1, kL_2) = -\dfrac{i\omega}{2k} M_1^a(kL_1), \\[2mm] J_{02}^s Y_{22}^{\Sigma ss}(kL_2, kL_2) + J_{01}^s Y_{21}^{\Sigma ss}(kL_2, kL_1) + J_{01}^a Y_{21}^{\Sigma sa}(kL_2, kL_1) = -\dfrac{i\omega}{2k} M_2^s(kL_2), \\[2mm] J_{02}^s Y_{22}^{\Sigma aa}(kL_2, kL_2) + J_{01}^s Y_{21}^{\Sigma as}(kL_2, kL_1) + J_{01}^a Y_{21}^{\Sigma aa}(kL_2, kL_1) = -\dfrac{i\omega}{2k} M_2^a(kL_2). \end{cases}$$
$$(4.22)$$

Here

$$Y_{mn}^{\Sigma \begin{Bmatrix} ss\ sa \\ as\ aa \end{Bmatrix}}(kL_m, kL_n)$$
$$= \frac{1}{2k} \int_{-L_m}^{L_m} f_m^{s,a}(s_m) \left[\left(\frac{d^2}{ds_m^2} + k^2 \right) \int_{-L_n}^{L_n} f_n^{s,a}(s_n') G_{s_n}^{\Sigma}(s_m, s_n') ds_n' \right] ds_m$$
$$(4.23)$$

are the eigen ($m = n$) and mutual ($m \neq n$) admittances of the slots;

$$M_m^{s,a}(kL_m) = \int_{-L_m}^{L_m} f_m^{s,a}(s_m) H_{0s_m}^{s,a}(s_m) ds_m \qquad (4.24)$$

are the partial magnetomotive forces. Due to (3.13) the functions $f_m^{s,a}(s_m)$ equal to:

$$f_m^s(s_m) = \cos k s_m \cos \gamma L_m - \cos k L_m \cos \gamma s_m,$$
$$f_m^a(s_m) = \sin k s_m \sin \gamma L_m - \sin k L_m \sin \gamma s_m. \qquad (4.25)$$

For two longitudinal slots in the common broad wall of rectangular waveguides, the S_{11} reflection and S_{12} transmission coefficients, the S_{13} and S_{14} transfer ones along the field and the $|S_\Sigma|^2$ power transfer coefficient have the form:

$$S_{11} = -\frac{4\pi k_c^2}{\omega a b \gamma H_0} \begin{Bmatrix} \cos \dfrac{\pi x_{01}}{a} \int_{-L_1}^{L_1} J_1(s_1) e^{-i\gamma s_1} ds_1 \\[3mm] + e^{-i\gamma z_0} \cos \dfrac{\pi x_{02}}{a} \int_{-L_2}^{L_2} J_2(s_2) e^{-i\gamma s_2} ds_2 \end{Bmatrix} e^{2i\gamma z},$$

$$S_{12} = 1 - \frac{4\pi k_c^2}{\omega a b \gamma H_0} \begin{Bmatrix} \cos \dfrac{\pi x_{01}}{a} \int_{-L_1}^{L_1} J_1(s_1) e^{i\gamma s_1} ds_1 \\[3mm] + e^{i\gamma z_0} \cos \dfrac{\pi x_{02}}{a} \int_{-L_2}^{L_2} J_2(s_2) e^{i\gamma s_2} ds_2 \end{Bmatrix},$$

$$S_{13} = S_{11}, \quad S_{14} = S_{12} - 1, \quad |S_\Sigma|^2 = |S_{13}|^2 + |S_{14}|^2 = 1 - |S_{11}|^2 - |S_{12}|^2.$$
$$(4.26)$$

In the case of electromagnetic coupling of two identical rectangular waveguides via coupling of longitudinal slots of the same lengths ($2L_1 = 2L_2 = 2L$) and widths ($d_1 = d_2 = d$), located at $z_0 = 0$, we obtain $Y_{mn}^{\Sigma sa, \Sigma as} = 0$, and the solution of the equations system (4.22) is sufficiently simplified. As a result we have ($Y_{mn}^{\Sigma ss, \Sigma aa} \rightarrow Y_{mn}^{\Sigma s, \Sigma a} = 2Y_{mn}^{is, ia}$):

$$J_{0m}^s = -\frac{i\omega}{2k^2} H_0 \tilde{J}_{0m}^s, \quad J_{0m}^a = -\frac{i\omega}{2k^2} H_0 i \tilde{J}_{0m}^a, \quad \begin{Bmatrix} m = 1, n = 2 \\ m = 2, n = 1 \end{Bmatrix},$$

$$\tilde{J}_{0m}^{s,a} = f^{s,a}(kL) \frac{\cos \dfrac{\pi x_{0m}}{a} Y_{nn}^{\Sigma s, \Sigma a}(kL) - \cos \dfrac{\pi x_{0n}}{a} Y_{mn}^{\Sigma s, \Sigma a}(kL)}{Y_{mm}^{\Sigma s, \Sigma a}(kL) Y_{nn}^{\Sigma s, \Sigma a}(kL) - Y_{mn}^{\Sigma s, \Sigma a}(kL) Y_{nm}^{\Sigma s, \Sigma a}(kL)}. \tag{4.27}$$

$$S_{11} = -\frac{2\pi k_c^2}{iab\gamma k^3} \left\{ \begin{array}{l} \cos \dfrac{\pi x_{01}}{a} [\tilde{J}_{01}^s f^s(kL) + \tilde{J}_{01}^a f^a(kL)] \\ + \cos \dfrac{\pi x_{02}}{a} [\tilde{J}_{02}^s f^s(kL) + \tilde{J}_{02}^a f^a(kL)] \end{array} \right\} e^{2i\gamma z},$$

$$S_{12} = 1 - \frac{2\pi k_c^2}{iab\gamma k^3} \left\{ \begin{array}{l} \cos \dfrac{\pi x_{01}}{a} [\tilde{J}_{01}^s f^s(kL) - \tilde{J}_{01}^a f^a(kL)] \\ + \cos \dfrac{\pi x_{02}}{a} [\tilde{J}_{02}^s f^s(kL) - \tilde{J}_{02}^a f^a(kL)] \end{array} \right\}, \tag{4.28}$$

where

$$f^s(kL) = 2\cos\gamma L_m \frac{\sin kL_m \cos\gamma L_m - (\gamma/k)\cos kL_m \sin\gamma L_m}{1 - (\gamma/k)^2}$$
$$- \cos kL_m \frac{\sin 2\gamma L_m + 2\gamma L_m}{2(\gamma/k)},$$

$$f^a(kL_m) = 2\sin\gamma L_m \frac{\cos kL_m \sin\gamma L_m - (\gamma/k)\sin kL_m \cos\gamma L_m}{1 - (\gamma/k)^2}$$
$$- \sin kL_m \frac{\sin 2\gamma L_m - 2\gamma L_m}{2(\gamma/k)}.$$

The expressions for the $Y_{mn}^{\begin{Bmatrix} ss\, sa \\ as\, aa \end{Bmatrix}}(kL_m, kL_n)$, $Y_{mn}^{\{s,a\}}(kL)$ admittances are given in Appendix C.

Figures 4.12 and 4.13 give the plots of the dependencies of the coupling coefficient $|S_\Sigma|^2 = |S_{13}|^2 + |S_{14}|^2$ from a slots electrical length at their different location relative to each other and the waveguide walls. As one can see, two closely located slots have sufficiently differing characteristics $|S_\Sigma|^2 = f(2L/\lambda)$ in dependence from their locations in a waveguide wall, and what's more, if one of the slots is on the axis line of a waveguide, two resonances located on the sides of resonance peaks for a single slot take place (Fig. 4.13). The explanation of this fact can be

Fig. 4.12 The coupling coefficient dependences from the relative length of the longitudinal slots in the common broad wall of two rectangular waveguides at: $a = 23.0$ mm, $b = 10.0$ mm, $d = a/15$, $\lambda/\lambda_c = 0.625$, $h = 2.0$ mm

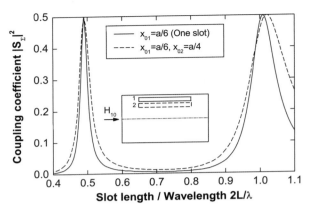

obtained with the help of the plots in Fig. 4.14, where we give the curves of change of the normalized values of current amplitudes and slots admittances in the range of $0.4 \leq 2L/\lambda \leq 0.5$, when main contribution to the currents is made by symmetrical components. At $x_{02} = a/2$ it results from (4.28) that energy characteristics of slots systems will be defined by the current in the first one of them, the amplitude of which equals (index "s" is omitted) due to (4.27):

$$\tilde{J}_{01} = \frac{\frac{1}{2} \cos \frac{\pi x_{01}}{a} f(kL)}{[\mathrm{Re}Y_{11}^i(kL) - \mathrm{Re}Y_c^i(kL)] + i\,\mathrm{Im}Y_{11}^i(kL)},$$

$$\mathrm{Re}Y_c^i(kL) = \frac{\mathrm{Re}Y_{12}^i(kL)\mathrm{Re}Y_{21}^i(kL)}{\mathrm{Re}Y_{22}^i(kL)}, \tag{4.29}$$

where we take into account that $Y_{mn}^i = \mathrm{Re}\,Y_{mn}^i + i\,\mathrm{Im}\,Y_{mn}^i$ and $\mathrm{Im}\,Y_{22}^i = \mathrm{Im}\,Y_{12}^i = \mathrm{Im}\,Y_{21}^i = 0$.

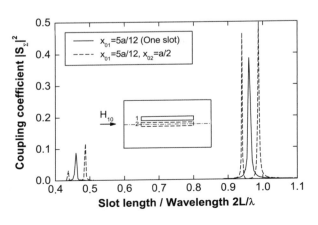

Fig. 4.13 The coupling coefficient dependences from the relative length of the longitudinal slots in the common broad wall of two rectangular waveguides at: $a = 23.0$ mm, $b = 10.0$ mm, $d = a/15$, $\lambda/\lambda_c = 0.625$, $h = 2.0$ mm

The analysis of formula (4.29) shows that in the case of the single slot ($Y_{12}^i = Y_{21}^i = 0$) a current amplitude reaches the maximum value if the $\mathrm{Re}Y_{11}^i(kL) = 0$ conditions are fulfilled, that is, when the reactive energy of all non-spreading modes of oscillations in the slot neighborhood equals to null (resonance). If we have another slot, located on the waveguide axis line, a $\mathrm{Re}Y_c^i(kL)$ value, proportional to $1/\mathrm{Re}Y_{22}^i(kL)$, is added to the $\mathrm{Re}Y_{11}^i(kL)$ value. What's more, it occurs because of smallness of the $\Delta x = |x_{01} - x_{02}|$ value, as is seen from the plots $\mathrm{Re}Y_{22}^i(kL) \cong \mathrm{Re}Y_{11}^i(kL)$, and $\mathrm{Re}Y_{12}^i(kL) = \mathrm{Re}Y_{21}^i(kL) \neq 0$. Thus, in this case the current amplitude is in inverse proportion to a function $F(x) = f(x) - \frac{c^2}{f(x)}$, the null equality of which, that is $\mathrm{Re}Y_{11}^i(kL) - \mathrm{Re}Y_c^i(kL) = 0$, is the resonance condition for the very slots couple. From the kind of $F(x)$ function it follows that the current amplitude is minimal at $\mathrm{Re}Y_{11}^i(kL) = 0$, and it has two maximums at $\mathrm{Re}Y_{11}^i \cong \pm\mathrm{Re}Y_{12}^i$, as it is shown in Fig. 4.14.

Let us note that in the following resonance ($0.9 \le 2L/\lambda \le 1.0$), all written above is valid, but the main contribution will be made into the current amplitude by the antisymmetrical component. From the plots in Fig. 4.15 it can be seen that the transmission coefficients into the second waveguide are equal between each other $|S_{13}| = |S_{14}|$ for the slots system in question, that is, such a structure does not have directed qualities at the given slots lengths. Both for a single slot and in the case of two slots, phase transition of the reflection coefficient via null ($\arg S_{11} = 0$) is the resonance condition at which the field amplitude, scattered by slots, sharply increases [4.5].

If a couple of slots are symmetrically located relative to the waveguide axis line (Fig. 4.16), then the decrease of the Δx distance between them leads to the goodness of resonance peaks (the interaction level of the slots with the exciting field decreases) and a significant shift of the slots resonance line into the region of "shortening" (in comparison with the adjusted slot $2L = n\lambda/2$, $n = 1,2,3...$), which corresponds to the character of change of the resonance line value of a single longitudinal slot in dependence of its location relative to the axis line of the broad waveguide wall [4.5].

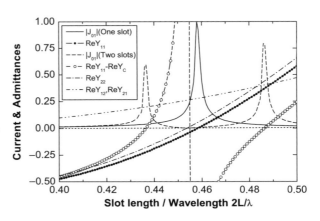

Fig. 4.14 The current and admittances dependences from the relative length of the longitudinal slots in the common broad wall of two rectangular waveguides at: $a = 23.0$ mm, $b = 10.0$ mm, $d = a/15$, $\lambda/\lambda_c = 0.625$, $h = 2.0$ mm

Fig. 4.15 The coupling coefficients dependences from the relative length of the longitudinal slots in the common broad wall of two rectangular waveguides at: $a = 23.0$ mm, $b = 10.0$ mm, $d = a/15$, $\lambda/\lambda_c = 0.625$, $h = 2.0$ mm

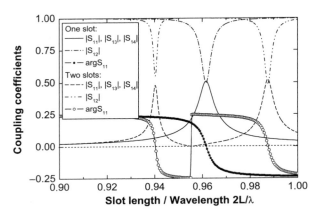

Fig. 4.16 The coupling coefficient dependence from the relative length of the longitudinal slots in the common broad wall of two rectangular waveguides at: $a = 23.0$ mm, $b = 10.0$ mm, $d = a/15$, $x_{02} = a - x_{01}$, $\lambda/\lambda_c = 0.625$, $h = 2.0$ mm

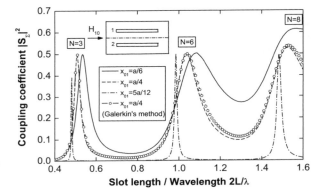

The comparison with Galerkin's method (Fig. 4.16, the curve —o—) proves the adequacy of the proposed basic functions to physically process this carefully considered waveguide-slotted structure.

4.4 Multi-Slot Coupling Through Symmetrical Transverse Slots in a Common Broad Wall of Waveguides

The system, depicted in Fig. 4.17, consists of N narrow ($\{d_n/(2L_n)\} << 1$, $\{d_n/\lambda\} << 1$, where $2L_n$, d_n are the length and width of the n-th slot) rectangular slots, located in the broad wall of thickness h of a rectangular waveguide with its cross section $\{a_1 \times b_1\}$ symmetrically about its longitudinal axis and radiating into the rectangular waveguide with cross section $\{a_2 \times b_2\}$.

After approximation of the current in the slots as

$$J_n^{i,e}(s_n) = J_{0n}^{i,e} f_n(s_n), \quad f_n(\pm L_n) = 0 \qquad (4.30)$$

Fig. 4.17 The multi-slot
system in the waveguides
common wall

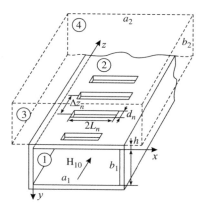

where $f_n(s_n)$ are some preset functions, and using the boundary conditions of continuity of tangential components of magnetic fields on both surfaces of each slot, we arrive at a SLAE in terms of unknown current amplitudes $J_{0n}^{i,e}$

$$\sum_{m=1}^{N}\sum_{n=1}^{N}(J_{0n}^{i}+J_{0n}^{e})Y_{mn}^{Wg^i,R,Wg^e}(kL_m,kL_n)=-\frac{i\omega}{2k}M_m^i(kL_m). \qquad (4.31)$$

where

$$Y_{mn}^{Wg^i,R,Wg^e}(kL_m,kL_n)$$

$$=\frac{1}{2k}\int_{-L_m}^{L_m}f_m(s_m)\left[\left(\frac{\mathrm{d}^2}{\mathrm{d}s_m^2}+k^2\right)\int_{-L_n}^{L_n}f_n(s_n')G_{s_m}^{Wg^i,R,Wg^e}(s_m,s_n')\mathrm{d}s_n'\right]\mathrm{d}s_m \qquad (4.32)$$

are the eigen ($m=n$) and mutual ($m\neq n$) slots admittances, respectively,

$$M_m^i(kL_m)=\int_{-L_m}^{L_m}f_m(s_m)H_{0s_m}(s_m)\,\mathrm{d}s_m \qquad (4.33)$$

are the magnetomotive forces, $G_{s_m}^{Wg^i,R,Wg^e}$ are the quasi-one-dimensional Green's magnetic function of the infinite (semi-infinite) rectangular waveguide (Wg), rectangular resonator (R), formed by the slot cavity, respectively, and H_{0s_m} are projections of the magnetic field from impressed sources on the axes of the slots.

The calculations and the comparison with the results obtained by other methods, showed that formulas (3.15) and (3.16), to define the slot effective width d_e, taking into account the finite thickness of the waveguides walls, become less accurate as the slots number increases ($N>2$). So, we consider one more electrodynamic volume, which is a rectangular resonator formed by the slot cavity. The $J_n^i(s_n)$ and $J_n^e(s_n)$ magnetic currents on the internal (index "i", region 1–2) and external (index "e", region 3–4) surfaces of the slots, correspondingly, will result from the SLAE (4.31) solution.

Provided that the system is excited by an H_{10}-type wave with amplitude H_0 propagating in the waveguide from the domain $z = -\infty$, let us take for $f_{m,n}(s_{m,n})$ the functions (4.16)

$$f_{m,n}(s_{m,n}) = \cos k s_{m,n} \cos k_c L_{m,n} - \cos k L_{m,n} \cos k_c s_{m,n}. \tag{4.34}$$

Having substituted (4.34) into (4.32) and (4.33), we find all coefficients of SLAE (4.31). Resolving this system permits us to find the energy characteristics of the waveguide-slotted structure under investigation.

The expressions for the S_{11} reflection and S_{12} transmission coefficients, the S_{13} and S_{14} transfer ones along the field and the $|S_\Sigma|^2$ power transfer coefficient of the structure under consideration can be written as

$$S_{11} = \frac{2\pi i \gamma_1}{a_1 b_1 k^3} \left\{ \sum_{n=1}^{N} J_{0n}^i F(k L_n) e^{-i\gamma_1 z_n} \right\} e^{2i\gamma_1 z},$$

$$S_{12} = 1 + \frac{2\pi i \gamma_1}{a_1 b_1 k^3} \left\{ \sum_{n=1}^{N} J_{0n}^i F(k L_n) e^{i\gamma_1 z_n} \right\},$$

$$S_{13} = \frac{2\pi i \gamma_2}{a_2 b_2 k^3} \left\{ \sum_{n=1}^{N} J_{0n}^e F(k L_n) e^{-i\gamma_2 z_n} \right\} e^{2i\gamma_2 z},$$

$$S_{14} = \frac{2\pi i \gamma_2}{a_2 b_2 k^3} \left\{ \sum_{n=1}^{N} J_{0n}^e F(k L_n) e^{i\gamma_2 z_n} \right\},$$

$$|S_\Sigma|^2 = |S_{13}|^2 + |S_{14}|^2, \tag{4.35}$$

where

$$F(k L_n) = 2 \cos k_c L_n \frac{\sin k L_n \cos k_c L_n - (k_c/k) \cos k L_n \sin k_c L_n}{1 - (k_c/k)^2}$$
$$- \cos k L_n \frac{\sin 2 k_c L_n + 2 k_c L_n}{2 k_c / k}.$$

The expressions for the $Y_{mn}^{Wg}(k L_m, k L_n)$, $Y_{mn}^R(k L_{m,n})$ admittances are given in Appendix C.

Figure 4.18 gives the dependences of the $|S_{11}|$, $|S_{12}|$, $|S_{13}|$, $|S_{14}|$ and $|S_\Sigma|^2$ coupling coefficients from the wavelength for the system consisting of 16 slots of equal length, the distance between which equals $z_{0m} = \lambda_g^{res}/4$, where λ_g^{res} is the waveguide wavelength, which corresponds to the λ_{res} resonant wavelength of the single slot (λ_{res}=33.7 mm for $2L$=16 mm). As it is seen from the plots, the power in this

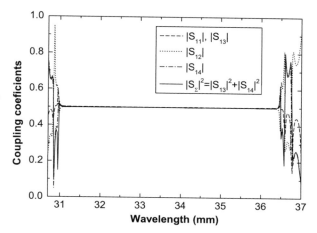

Fig. 4.18 The coupling coefficients dependences from the wavelength for the system of 16 transverse slots in the common broad wall of two identical rectangular waveguides at: $a = 23.0$ mm, $b = 10.0$ mm, $d_n = 1.6$ mm, $2L_n = 16.0$ mm, $\Delta z_n = 12.4$ mm, $h = 0.2$ mm

case, entering the first shoulder of the main waveguide (a region 1) is divided into 4 equal parts in a sufficiently wide range of wavelengths ($\Delta\lambda/\lambda_{res} = 0.15$).

However, the quantitative ratios of the coupling coefficients relative to each other can be distributed differently in other parts of the H_{10}-wave band. Thus, for example, the region, where the whole power of the incident wave propagates along the main waveguide (from region 1 to region 2) and does not enter into the second waveguide (Fig. 4.19), exists. On the contrary, the directional coupling takes place in definite parts of the one-mode band: from region 1 into region 4 (Fig. 4.20); moreover the $|S_{14}|$ coefficient value can amount to the value 1.0 (Fig. 4.20b).

Note that here we take into account full interaction between all slots for a finite thickness of waveguides walls, and the calculation time is far less then when Galerkin's or the finite elements methods are used.

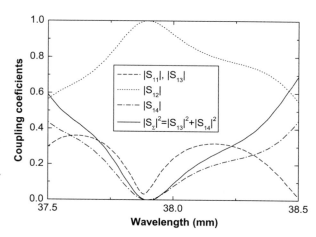

Fig. 4.19 The coupling coefficients dependences from the wavelength for the system of 16 transverse slots in the common broad wall of two identical rectangular waveguides at: $a = 23.0$ mm, $b = 10.0$ mm, $d_n = 1.6$ mm, $2L_n = 16.0$ mm, $\Delta z_n = 12.4$ mm, $h = 0.2$ mm

Fig. 4.20 The coupling coefficients dependences from the wavelength for the system of 16 transverse slots in the common broad wall of two identical rectangular waveguides at: $a = 23.0$ mm, $b = 10.0$ mm, $d_n = 1.6$ mm, $2L_n = 16.0$ mm, $\Delta z_n = 12.4$ mm, $h = 0.2$ mm

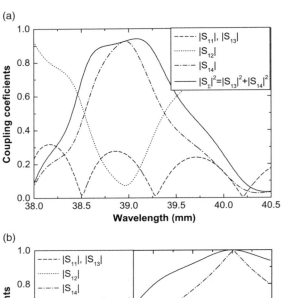

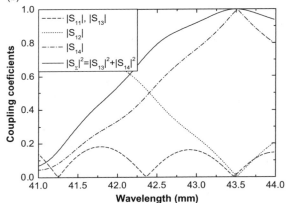

References

4.1. Whetten, F.L., Balanis, C.A.: Meandering long slot leaky-wave waveguide antennas. IEEE Trans. Antennas and Propagat. **AP-39**, 1553–1560 (1991)

4.2. Josefsson, L.G.: Analysis of longitudinal slots in rectangular waveguides. IEEE Trans. Antennas and Propagat. **AP-35**, 1351–1357 (1987)

4.3. Katrich, V.A., Lyashchenko, V.A., Berdnik, S.L.: Electrically long waveguide-slot antennas with optimal radiation and directivity characteristics. Radioelectronics and Communications Systems. **46**, 36–42 (2003)

4.4. Jia, H., Yoshitomi, K., Yasumoto, K.: Rigorous and fast convergent analysis of a rectangular waveguide coupler slotted in common wall. Progress In Electromagnetics Research. **PIER 46**, 245–264 (2004)

4.5. Katrich, V.A., Nesterenko, M.V.: The near-zone field and resonant frequencies of narrow longitudinal slots in the broad face of a rectangular waveguide. Telecommunications and Radio Engineering. **60**, 125–131 (2003)

Chapter 5
Resonant Iris with the Slot Arbitrary Oriented in a Rectangular Waveguide

With the development of microwave technology, the resonant irises, which have a length and width less than the sizes of a waveguide cross section, are the integral part of the elementary basis of different devices, for example, band-pass and band-rejection filters [5.1,5.2,5.3], transformers and diaphragmatic resonant joints of rectangular waveguides [5.4, 5.5], etc. Starting with the classical monograph written by Levin L. [5.6], a considerable number of publications were devoted to the investigation of electrodynamic characteristics of a resonant iris directly as an elementary cell of complex waveguides units. In these papers the irises of infinite small and finite thicknesses, the slot axes of which are parallel to the broad walls of a rectangular waveguide (coordinate irises), were analyzed by different methods (both analytical and numerical). The investigation of the rectangular slot arbitrary located in the plane of a waveguide was evidently made in [5.7] by the numerical method of moments. However, the calculated and experimental values given in [5.7] correspond to the non-resonant apertures (the coordinate iris is an exception) and do not answer the question how angular deflection of the slot in the plane of the cross section of a waveguide influences the electrodynamic characteristics of a resonant iris. These investigations were made in [5.8], where the task of an oblique iris of the finite thickness in a rectangular waveguide has been solved by means of the rigorous method of the generalized matrices of scattering (matrix operators), and some calculated results are given for the reflection coefficient and goodness of hollow and filled by dielectric irises in a rectangular and square waveguides.

This chapter represents the problem solution of electromagnetic waves scattering on the resonant iris of the finite thickness with the arbitrary oriented slot in the plane of the cross section of a rectangular waveguide. The influence of geometric parameters of such a structure on its electrodynamic characteristics has been analyzed in detail on the basis of the obtained approximate analytical problem solution.

5.1 Problem Formulation

Let the resonant iris, the slot of which is arbitrary oriented in the plane of the cross section of a waveguide (Fig. 5.1) be located in the region $0 \leq z \leq h$ of an infinite rectangular waveguide by the section $\{a \times b\}$. The impressed sources of harmonic

M.V. Nesterenko et al., *Analytical and Hybrid Methods in the Theory of Slot-Hole Coupling of Electrodynamic Volumes*, DOI: 10.1007/978-0-387-76362-0_5,
© Springer Science+Business Media, LLC, 2008

Fig. 5.1 The problem
formulation and the symbols
used

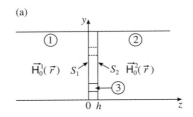

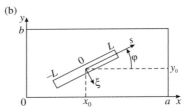

electromagnetic fields set by the magnetic intensities $\vec{H}_0^1(\vec{r})$ and $\vec{H}_0^2(\vec{r})$ (where $\vec{r}$ is the radius-vector in Cartesian coordinate system (x, y, z)) are situated in volumes 1 and 2, representing semi-infinite rectangular waveguides. Volume 3 (a slot cavity) is limited by the surfaces S_1 and S_2 and is free from the impressed sources.

Let us formulate the boundary conditions of continuity of tangential components of full magnetic fields on the apertures S_1 and S_2 in order to solve the problem of electromagnetic waves scattering on the iris in question:

$$
\begin{cases}
\vec{H}_{0\tau}^1(\vec{r}) + \vec{H}_\tau^1[\vec{r}, \vec{J}_1(\vec{r})] = \vec{H}_\tau^3[\vec{r}, \vec{J}_1(\vec{r})] + \vec{H}_\tau^3[\vec{r}, \vec{J}_2(\vec{r})]\Big|_{\text{on } S_1}, \\
\vec{H}_\tau^3[\vec{r}, \vec{J}_1(\vec{r})] + \vec{H}_\tau^3[\vec{r}, \vec{J}_2(\vec{r})] = \vec{H}_{0\tau}^2(\vec{r}) + \vec{H}_\tau^2[\vec{r}, \vec{J}_2(\vec{r})]\Big|_{\text{on } S_2},
\end{cases}
\tag{5.1}
$$

where $\vec{J}_1(\vec{r})$ and $\vec{J}_2(\vec{r})$—the unknown equivalent surface magnetic currents on the holes S_1 and S_2 at their conditional metallization; $\vec{H}_\tau^1$, $\vec{H}_\tau^2$, $\vec{H}_\tau^3$—the tangential (respectively to the iris plane) components of the scattering magnetic fields in volumes 1, 2 and 3, correspondingly.

Further we suppose that the slot of the iris is narrow; that is, its length $2L$ and its width d satisfy the inequations

$$
\frac{d}{2L} \ll 1, \quad \frac{d}{\lambda} \ll 1.
\tag{5.2}
$$

Then the surface magnetic currents on S_1 and S_2 can be represented in the following form:

$$
\vec{J}_1(s_1, \xi_1) = \vec{e}_{s_1} J_1(s_1)\chi(\xi_1), \quad \vec{J}_2(s_2, \xi_2) = \vec{e}_{s_2} J_2(s_2)\chi(\xi_2),
\tag{5.3a}
$$

$$
J_1(\pm L) = 0, \quad J_2(\pm L) = 0.
\tag{5.3b}
$$

Here $\vec{e}_{s_1}$ and $\vec{e}_{s_2}$ are the orts of the local coordinates (s, ξ) coupled with the slot (Fig. 5.1b), and $\chi(\xi)$ is the function, taking account the electric field behavior on the edges of the slot cavity [5.9].

Function $\chi(\xi)$, satisfying the normalization condition:

$$\int_{-d/2}^{d/2} \chi(\xi)d\xi = 1. \tag{5.4}$$

Particularly, at $h = 0$ (the infinite thin iris) $\chi(\xi)$ has the form:

$$\chi(\xi) = \frac{1/\pi}{\sqrt{(d/2)^2 - \xi^2}}. \tag{5.5}$$

In the case where $h \neq 0$ (the iris of the finite thickness and each of the edges of the slot cavity represented a perfectly conductive rectangular wedge) $\chi(\xi)$ has the form:

$$\chi(\xi) = \frac{\Gamma(7/6)/\Gamma(2/3)}{\sqrt{\pi} \ (d/2) \ \sqrt[3]{1 - (2\xi/d)^2}}, \tag{5.6}$$

where $\Gamma(x)$ is the Gamma-function.

It is shown in [3.1] that if $H^1_{0s_1}(s_1) + H^2_{0s_2}(s_2) \neq 0$ (where $H^1_{0s_1}(s_1)$ and $H^2_{0s_2}(s_2)$ are the projections of the magnetic fields of the impressed sources to the longitudinal slot axes) and $(h/\lambda) \ll 1$, one can suggest that $J_1(s_1) \approx J_2(s_2) = J(s)$ with the accuracy to the values of the $\{(hd)/\lambda^2\}$ order under the condition that volume 3 is a rectangular resonator (volumes 1 and 2 are arbitrary). Then the problem of the d slot width in the wall of the h finite thickness is reduced to the problem of the slot with "the equivalent" width d_e in the infinite thin wall after substitution (5.3a) into the equations system (5.1), taking into account (5.4)–(5.6):

$$H^1_{0s}(s) + H^1_s[s, J(s)] = H^2_{0s}(s) + H^2_s[s, J(s)] \Big|_{z=0}. \tag{5.7}$$

For all that, d_e is coupled with d in the following way [3.1, 3.2]:

$$d_e = d \frac{K}{E(K)}, \tag{5.8}$$

where $E(K)$ is the complete elliptic integral of the second kind, and K is the integral modulus. We obtain suitable calculated ratios from (5.8) in the extreme cases as is shown in Sect. 3.3 (formulas 3.15 and 3.16).

Taking into account all the above-mentioned, and supposing that the impressed field in the second volume equals null (index "1" is omitted), we get the integral–differential equation concerning the $J(s)$ magnetic current in the slot from (5.7):

$$\left(\frac{d^2}{ds^2} + k^2\right) \int\limits_{-L}^{L} J(s')[G_s^1(d_e; s, s') + G_s^2(d_e; s, s')] \, ds' = -i\omega H_{0s}(s). \quad (5.9)$$

Here the functions $G_s^{1,2}(d_e; s, s')$ are the s-components of the quasi-one-dimensional $|\xi - \xi'| \approx d_e/4$ magnetic dyadic Green's function $\hat{G}^m(\vec{r}, \vec{r}')$ for the Hertz's vector potentials of the first and second volumes, correspondingly (see Appendix A).

5.2 Solution of the Equation for a Magnetic Current

In Chap. 3 we have obtained the asymptotic solution of the integral–differential equation (5.9) for the magnetic current in the narrow slot, coupling two arbitrary electrodynamic volumes with the help of the averaging method:

$$J^{s,a}(s) = \overline{A}(-L) \cos ks + \overline{B}(-L) \sin ks$$

$$+ \alpha \int\limits_{-L}^{s} \left\{ \frac{i\omega}{k} H_{0s}^{s,a}(s') + \overline{F}_N^{s,a}[s', \overline{A}, \overline{B}] \right\} \sin k(s - s') ds'. \quad (5.10)$$

Here $\alpha = \frac{1}{8\ln[d_e/(8L)]}$ is the natural small parameter of the problem.

Supposing that the wave of the main mode H_{10} with the H_0 amplitude propagates from region $z = -\infty$ in the first waveguide, we have:

$$H_{0s}(s) = 2H_0 \cos\varphi \left[\sin\frac{\pi x_0}{a} \cos\frac{\pi(s\cos\varphi)}{a} + \cos\frac{\pi x_0}{a} \sin\frac{\pi(s\cos\varphi)}{a} \right]. \quad (5.11)$$

Then the current in the slot has the following form, taking into account its symmetrical and antisymmetrical components (relatively to the slot center $s = 0$):

$$J(s) = J_0 f(s) = -\alpha 2 H_0 \cos\varphi \frac{2i\omega/k^2}{[1 - (\tilde{k}/k)^2][\sin 2kL + \alpha 2W_0(kd_e, 2kL)]}$$

$$\times \left\{ \begin{array}{l} \sin\dfrac{\pi x_0}{a} \sin kL(\cos ks \cos \tilde{k}L - \cos kL \cos \tilde{k}s) \\[2mm] + \cos\dfrac{\pi x_0}{a} \cos kL(\sin ks \sin \tilde{k}L - \sin kL \sin \tilde{k}s) \end{array} \right\}, \quad (5.12)$$

where J_0 is the amplitude, $f(s)$ is the current distribution function, $\tilde{k} = \frac{\pi}{a}\cos\varphi$, $W_0(kd_e, 2kL)$ is the slot own field function, defined by the corresponding components of the magnetic dyadic Green's function of a semi-infinite rectangular waveguide.

It follows from the analysis of the current expression (5.12) that in the case of a resonant slot ($kL \approx \pi/2$) its antisymmetrical component is always small in comparison with a symmetrical one because $|\cos \frac{\pi x_0}{a} \cos kL| << |\sin \frac{\pi x_0}{a} \sin kL|$ in the frequencies range of the H_{10} wave mode of a standard rectangular waveguide. So, to investigate the influence of the angle of rotation φ on the iris characteristics we assume in formula (5.12) $x_0 = a/2$, $y_0 = b/2$. Then, taking into account that the ratio is performed,

$$W_0(kd_e, 2kL) = 2 \sin kL W_\varphi(kd_e, kL), \qquad (5.13)$$

the expressions for the current and the own field functions of the slot are considerably simplified:

$$J(s) = -\alpha H_0 \cos \varphi \left(\frac{2i\omega}{k^2} \right) \frac{\left\{ \cos ks \cos \frac{\pi}{a}(L \cos \varphi) - \cos kL \cos \frac{\pi}{a}(s \cos \varphi) \right\}}{\left[1 - \left(\frac{\pi}{ka} \cos \varphi \right)^2 \right] [\cos kL + \alpha 2 W_\varphi(kd_e, kL)]}. \qquad (5.14)$$

We note, that if $x_0 = a/2$ and $y_0 \neq b/2$, and $\varphi \neq 0$, then the addendums, proportional to $\cos kL$, will be present in the own field function of the slot $W_0(kd_e, 2kL)$; that is, the ratio (5.13) not will be performed. And for all that, both "symmetrical" and "antisymmetrical" components are not defined by the slot excitation (in $H_{0s}(s)$ only a symmetrical component is present), but its location relative to the waveguide walls, and they influence the near field of the iris.

The formula for the current (5.14) fully defines the S_{11} reflection coefficient and S_{12} transmission coefficient of the fundamental wave in the considered slot iris:

$$S_{11} = (1 + S_{12}) \, e^{2i\gamma z}, \quad S_{12} = -\alpha \frac{16\pi \gamma \cos^2 \varphi \, f_\varphi(\tilde{k}L)}{i a b k^3 [1 - (\tilde{k}/k)^2][\cos kL + \alpha 2 W_\varphi(kd_e, kL)]}. \qquad (5.15)$$

Here

$$f_\varphi(\tilde{k}L) = 2 \cos \tilde{k}L \frac{\sin kL \cos \tilde{k}L - (\tilde{k}/k) \cos kL \sin \tilde{k}L}{1 - (\tilde{k}/k)^2} - \cos kL \frac{\sin 2\tilde{k}L + 2\tilde{k}L}{2(\tilde{k}/k)}.$$

The dispersive equation, defining its resonant frequencies, on which the equality of period average values of electrical and magnetic energies of the near fields takes place, follows from the null equality condition of the imaginary part of the reflection coefficient from the iris $\text{Im} S_{11} = 0$:

$$\cos(kL)_{res} + \alpha 2 \text{Re} \left\{ W_\varphi (kd_e, (kL)_{res}) \right\} = 0, \qquad (5.16)$$

at the same time the current distribution (5.14) has the maximum amplitude and null phase.

Let us define the approximate solution of equation (5.16), expanding the unknown value $(kL)_{res}$ into the series due to the small α parameter degrees:

$$(kL)_{res} = (kL)_0 + \alpha(kL)_1 + \alpha^2(kL)_2 + ... \tag{5.17}$$

If we substitute (5.17) into (5.16) and make equal the addendums at the same α degrees with the accuracy to the α^2 order terms, we get:

$$(kL)_{res} \approx \frac{\pi}{2} + \alpha 2\mathrm{Re}\left\{W_\varphi\left(\frac{\pi d_e}{2L}, \frac{\pi}{2}\right)\right\}, \tag{5.18}$$

where $\mathrm{Re}\left\{W_\varphi\right\}$ is the real part of the W_φ.

If $x_0 = a/2$, $\varphi = 0$, and y_0 is arbitrary (the coordinate symmetrical iris), then the expressions for the current and reflection coefficient have the form:

$$J(s) = -\alpha H_0\left(\frac{2i\omega}{\gamma^2}\right)\frac{\left(\cos ks\cos\frac{\pi}{a}L - \cos kL\cos\frac{\pi}{a}s\right)}{\cos kL + \alpha 2W(kd_e, kL)}, \tag{5.19}$$

$$S_{11} = \left\{1 - \alpha\frac{16\pi\, f\left(kL, \frac{\pi L}{a}\right)}{iabk\gamma[\cos kL + \alpha 2W(kd_e, kL)]}\right\}e^{2i\gamma z}, \tag{5.20}$$

where

$$f\left(kL, \frac{\pi L}{a}\right) = 2\cos\frac{\pi L}{a}\frac{\sin kL\cos\frac{\pi L}{a} - \left(\frac{\pi}{ka}\right)\cos kL\sin\frac{\pi L}{a}}{1 - (\pi/ka)^2}$$
$$- \cos kL\frac{\sin\frac{2\pi L}{a} + \frac{2\pi L}{a}}{(2\pi/ka)}.$$

The own field function of the slot $W(kd_e, kL)$ equals at $kL = \pi/2$:

$$W\left(\frac{\pi d_e}{2L}, \frac{\pi}{2}\right) = \frac{\pi^2}{abL}\sum_{m=1,3...}^{\infty}\sum_{n=0}^{\infty}\frac{\varepsilon_n\cos^2 k_x L}{k_z[(\pi/(2L))^2 - k_x^2]}\cos k_y y_0\cos k_y\left(y_0 + \frac{d_e}{4}\right). \tag{5.21}$$

Here $\varepsilon_n = 1$ at $n = 0$; $\varepsilon_n = 2$ at $n \neq 0$; $k_x = m\pi/a$, $k_y = n\pi/b$, $k_z = \sqrt{k_x^2 + k_y^2 - (\pi/(2L))^2}$ (m, n are integers).

Formula (5.21) can result from expressions not containing double series (see Appendix D). Then we get the formula for the resonant wavelength λ_{res} of the symmetrical ($x_0 = a/2$) iris with the slot length $2L$, the width d and the thickness h at its arbitrary location relatively to the $y_0 = b/2$ line in the waveguide of the $\{a \times b\}$ cross section from:

$$\frac{\lambda_{res}}{\lambda_c} = \frac{2L/a}{1 + \alpha \left(\dfrac{2}{\pi}\right) \mathrm{Re}\,W \left(\dfrac{\pi d_e}{2L}, \dfrac{\pi}{2}\right)}, \tag{5.22}$$

where λ_c is the cut-off H_{10} wavelength,

$$\mathrm{Re}\,W \left(\frac{\pi d_e}{2L}, \frac{\pi}{2}\right) \cong 2\pi \left\{ \frac{4\cos^2 \dfrac{\pi L}{a}}{\gamma_{10}^2 aL} \left[\frac{2\pi \cos^2 \dfrac{\pi y_0}{b}}{k_{11} b} - 2\cos^2 \frac{\pi y_0}{b} \right. \right.$$
$$\left. + \left(\frac{\gamma_{10} b}{9}\right)^2 \cos^2 \frac{2\pi y_0}{b} - \ln \left(\frac{\pi d_e}{2b} \sin \frac{\pi y_0}{b}\right) \right]$$

$$- \frac{4\cos^2 \dfrac{3\pi L}{a}}{k_{30}^2 aL} \left[K_0 \left(k_{30} \frac{d_e}{4}\right) + K_0(2k_{30} y_0) \right] + \left(\ln \frac{16L}{d_e} - 1\right)$$

$$- \ln \frac{1 - L/a}{1 + L/a} + \ln \frac{1 - 2L/a}{1 + 2L/a} + \ln \frac{1 - 2L/(3a)}{1 + 2L/(3a)} \tag{5.23}$$

$$- \frac{a}{2L} \left[\ln \left(1 - \left(\frac{2L}{a}\right)^2\right) - 2\ln \left(1 - \left(\frac{L}{a}\right)^2\right) + 3\ln \left(1 - \left(\frac{2L}{3a}\right)^2\right) - \left(\frac{L}{a}\right)^2 \right]$$

$$- \frac{4a}{\pi^2 L} \left[K_0 \left(\frac{\pi d_e}{4a}\right) \sin^2 \frac{\pi L}{a} + \frac{1}{9} K_0 \left(\frac{3\pi d_e}{4a}\right) \sin^2 \frac{3\pi L}{a} \right.$$
$$\left. \left. + \sum_{m=5,7\ldots}^{\infty} K_0 \left(\frac{m\pi d_e}{4a}\right) \Big/ m^2 \right] \right\}.$$

In (5.23) we have the symbols: $\gamma_{10} = \sqrt{(\pi/(2L))^2 - (\pi/a)^2}$, $k_{11} = \sqrt{(\pi/a)^2 + (\pi/b)^2 - (\pi/(2L))^2}$, $k_{30} = \sqrt{(3\pi/a)^2 - (\pi/(2L))^2}$, $d_e = d\exp(-\pi h/(2d))$, $K_0(x)$ is the McDonald's function. As $K_0(x)$ decreases rapidly at the x increase, then it is sufficient to take some first terms when summing up the rest series in (5.23).

We note, that in the case where $\{d_e/(2L)\} \to 0$, the expression (5.22) transits into the classical Slatter's formula for the symmetrical resonant "window" in a rectangular waveguide, given, for example, in [5.2]. There we have obtained more accurate expression for the λ_{res} slot iris; however, it is just only when the conditions $y_0 = b/2$ and $h = 0$ are fulfilled.

The expressions for $W_0(kd_e, 2kL)$, $W_\varphi(kd_e, kL)$, $W(kd_e, kL)$, functions are represented in Appendix B.

5.3 Numerical Results

Figure 5.2 represents the dependencies of the reflection coefficient $|S_{11}|^2$ from the wavelength λ of the H_{10}-wave mode, calculated due to formula (5.15) (solid curves) for the iris, which has an angle of $30°$ (Fig. 5.2a) between the s slot axis and the $\{0x\}$ waveguide axis, and for the coordinate iris (Fig. 5.2b, $\varphi = 0°$). Here we also marked the experimental value (the circles) and the numerical results (the dotted curves) obtained with the use of the "CST Microwave Studio" program. The comparison of

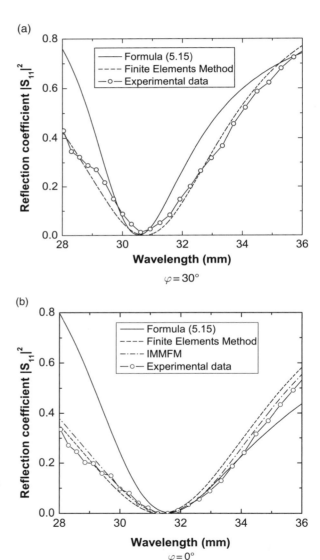

Fig. 5.2 The reflection coefficient dependences from the wavelength for an iris in rectangular waveguide at: $a = 23.0$ mm, $b = 10.0$ mm, $d = 1.5$ mm, $2L = 16.0$ mm, $x_0 = a/2$, $y_0 = b/2$, $h = 2.0$ mm

the curves shows, that the calculations, made due to the approximate formula (5.15), give satisfactory results near the resonance λ_{res} wavelength of the iris (the $|S_{11}|^2$ minimal values region), and they qualitatively describe the characteristics change (the structure Q-factor increases at the slot turn) in other parts of the range. It is explained that the asymptotic solution (5.10) of the integral–differential equation (5.9) was obtained in Chap. 3 by means of the averaging method in the first approximation along the α small parameter; that is with the accuracy up to the terms of the α^2 order. To our minds, it is not worthwhile to define further approximations for the current by this method because of the inconvenience of the obtained expressions in the case of non-coordinate irises or some slots system. However, the obtained expressions for the $f(s)$ distribution functions of the magnetic current in the formulas (5.12), (5.14) and (5.19) can be used as basic ones to solve equation (5.9) by means of the induced magnetomotive forces method (IMMFM) (see Chap. 4).

As example Fig. 5.2b gives the $|S_{11}|^2$ calculated values, obtained by the IMMFM with the use of the approximating functions for the current in the coordinate slot of the type (5.19) (stroke-dotted curve).

Let us analyze influence of the angle of turn and the slot shift in the plane of the waveguide cross section on a wave resonance length (frequency) of the considered iris. Figure 5.3 represents the dependencies of the arg S_{11} reflection coefficient phase (normalized on π) from the wavelength for different angles of inclination φ of the slot. As can be seen λ_{res} (the resonance is defined by the equality arg $S_{11} = 0$) shifts first to the short-wave part of the H_{10} range of the wave (slot "lengthening" in comparison with "adjusted" one, when $2L = \lambda_{res}/2$, where λ_{res}=32.0 mm) at the angle φ increase. Then the resonance wavelength increases at some values of the angles ($\varphi \approx 30°$). This conformity is illustrated in Fig. 5.4, where we give the values of the resonance wavelength of the λ_{res}/λ_c angular irises (they are normalized on the λ_c) at different slot lengths and for two values of the iris thickness: $h = 0.1$ mm, 1.0 mm, calculated according to formula (5.18).

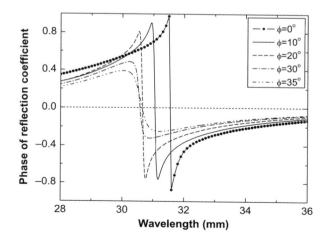

Fig. 5.3 The phase of reflection coefficient (normalized on π) dependences from the wavelength for an iris in rectangular waveguide at: $a = 23.0$ mm, $b = 10.0$ mm, $d = 1.5$ mm, $2L = 16.0$ mm, $x_0 = a/2$, $y_0 = b/2$, $h = 2.0$ mm

Fig. 5.4 The relative resonant wavelength dependences from the rotation angle for an iris in rectangular waveguide at: $a = 23.0$ mm, $b = 10.0$ mm, $d = 1.5$ mm, $2L = 16.0$ mm, $x_0 = a/2$, $y_0 = b/2$; solid curves – $h = 1.0$ mm, dotted curves – $h = 0.1$ mm, stroke-dotted curves – $2L = \lambda_{res}/2$

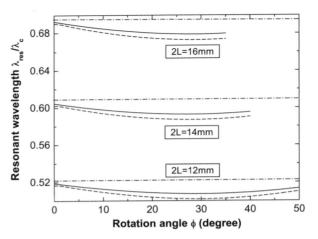

The resonant wavelength increases in the case of the coordinate iris shift $\varphi = 0°$ from the center ($y_0/b = 0.5$) of the waveguide (the calculations are made according to formula (5.22) for Fig. 5.5). In this case "lengthening" of the slot transits into its "shortening," reaching the $2L = \lambda_{res}/2$ value at different values of y_0/b for different slot lengths. We also note that influence of the h iris thickness on λ_{res} turns out to be more sufficient in the case of the slot inclination than at its shift to the waveguide broad wall.

Different behavior of the resonant curves in Fig. 5.4 and Fig. 5.5 can be explained in the following way. The equality conditions of period average values of electrical and magnetic energies (resonance) of the iris near fields (defined by the functions W_0, W_φ and W of the own field of the slot) can be reached by two methods: slot geometrical sizes change relative to the wavelength or its location change relative to the waveguide walls at the fixed electrical sizes. So, additional oscillations of the

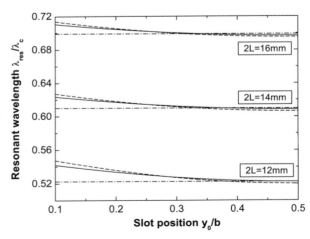

Fig. 5.5 The relative resonant wavelength dependences from the slot position for an iris in rectangular waveguide at: $a = 23.0$ mm, $b = 10.0$ mm, $d = 1.5$ mm, $x_0 = a/2$, $\varphi = 0°$; solid curves – $h = 1.0$ mm, dotted curves – $h = 0.1$ mm, stroke-dotted curves – $2L = \lambda_{res}/2$

Table 5.1 The resonant frequencies (in GHz) of a symmetrical coordinate irises in the rectangular waveguide at: $a = 22.86$ mm, $b = 10.16$ mm, $x_0 = a/2$, $y_0 = b/2$, $\varphi = 0°$, $h = 0.1$ mm

$2L \times d$, mm	Computation [5.7]	Experimental data [5.7]	Computation, formula (5.22)
16.9×0.9	8.87	8.84	8.84
14.8×0.5	10.22	10.20	10.13
12.9×0.9	11.62	11.65	11.66

E-mode ($E_{m,2n-1}$) appear in the near field (the own field function of the W_0 takes this into account) at the slot rotation to the narrow wall of a waveguide relative to the normal, and the H-mode ($H_{m,2n-1}$) oscillations appear at shift towards the waveguide broad wall; this is stipulated by the difference in redistribution of reactive energy and thus in the character of the resonant curves.

In order to estimate the accuracy of the obtained analytical formulas for $(kL)_{res}$ we have made the comparison with the numerical and experimental values due to the f_{res} frequencies of symmetrical coordinate irises ($x_0 = a/2$, $y_0 = b/2$, $\varphi = 0°$) of different slot lengths and widths, given in [5.7] and Table 5.1.

References

5.1. Southworth, G.C.: Principles and Applications of Waveguide Transmission. New York (1950)
5.2. Chen, T.S.: Waveguide resonant-iris filters with very wide passband and stopbands. Int. J. Electronics. **21**, 401–424 (1966)
5.3. Bornemann, J., Vahldieck, R.: Characterization of a class of waveguide discontinuities using a modified TE_{mn}^x mode approach. IEEE Trans. Microwave Theory and Tech. **MTT-38**, 1816–1822 (1990)
5.4. Patzelt, H., Arndt, F.: Double-plane steps in rectangular waveguides and their application for transformers, irises, and filters. IEEE Trans. Microwave Theory and Tech. **MTT-30**, 771–776 (1982)
5.5. Beyer, R., Arndt, F.: Efficient modal analysis of waveguide filters including the orthogonal mode coupling elements by an MM/FE method. IEEE Microwave and Guided Wave Lett. **5**, 9–11 (1995)
5.6. Lewin, L.: Advanced Theory of Waveguides. Iliffe & Sons, London (1951)
5.7. Yang, R., Omar, A.S.: Analysis of this inclined rectangular aperture with arbitrary location in rectangular waveguide. IEEE Trans. Microwave Theory and Tech. **MTT-41**, 1461–1463 (1993)
5.8. Rud, L.A.: Axially rotated step junction of rectangular waveguides and resonant diaphragms based thereupon. Telecommunications and Radio Engineering. **55**, 17–26 (2001)
5.9. Mittra, R., Lee, S.W.: Analytical Techniques in the Theory of Guided Waves. The Macmillan Company, NY (1971)
5.10. Gradshteyn, I.S., Ryzhik, I.M.: Table of Integrals, Series, and Products. Academic Press, NY (1980)

Chapter 6
Stepped Junction of the Two Rectangular Waveguides with the Impedance Slotted Iris

In the previous chapter we considered the problem of the resonant slot iris in the infinite rectangular waveguide in the case where all surfaces of waveguide elements were supposed to be perfectly conductive ones.

Thin-film coatings of materials with variable electrophysical properties can be used for expansion of functional capabilities of waveguide devices and realization of non-mechanical control of their characteristics by means of external field-effect or other actions. In particular, the coatings can be applied directly on iris surfaces.

Apparently, that experimental development of waveguide devices containing elements with thin-film coatings is a very laborious process. In this connection evolution of the mathematical models of devices and elements of such kind is an important problem for practical applications. Here modeling is usually based on solutions of boundary–value problems in which the approximation of a distributed surface impedance is used [6.1, 6.2]. On the one hand, this simplifies the problem solution, and on the other hand, this allows one to generalize using an obtained mathematical model.

The purpose of this chapter is the development of an approximate analytical solution of the diffraction problem of H_{10}-wave by an impedance slot iris of finite thickness situated in the plane of stepped rectangular waveguide couplings that differ in dimensions of cross sections. In this case generally both surfaces of an iris will be assumed to be impedance.

6.1 Problem Formulation

Let impressed sources of the electromagnetic field $\vec{H}_0^{in}(\vec{r}_{in})$ (where $\vec{r}_{in}$ is the radius-vector in the local Cartesian system (x_{in}, y_{in}, z_{in}) related to a waveguide section as is shown in Fig. 6.1) be situated in an internal region (denoted by the index "in") of a semi-infinite rectangular waveguide of cross section $\{a_{in} \times b_{in}\}$ with perfectly conductive side walls.

A narrow rectilinear slot of the length $2L$ and width d $((d/2L)<<1, (d/\lambda)<<1)$ is made in the end wall of the waveguide section of the thickness h $((h/\lambda)<<1)$. The long axis of the slot is parallel to the axis x_{in}, the center is at the point with

M.V. Nesterenko et al., *Analytical and Hybrid Methods in the Theory of Slot-Hole Coupling of Electrodynamic Volumes*, DOI: 10.1007/978-0-387-76362-0_6, © Springer Science+Business Media, LLC, 2008

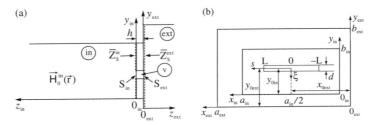

Fig. 6.1 The problem formulation and the symbols used

coordinates $(x_{0in}, y_{0in}, 0_{in})$, and $x_{0in} = a_{in}/2$, i.e., the slot is situated symmetrically with respect to the axis line of the broad wall of the waveguide section. There are no impressed sources in the external region, which is denoted by the index "*ext*". This region is a semi-infinite rectangular waveguide of cross section $\{a_{ext} \times b_{ext}\}$ with perfectly conductive side walls. The center of coupling aperture with the waveguide section "*in*" will be determined by coordinates $x_{0ext}, y_{0ext}, 0_{ext}$ in the Cartesian system $x_{ext}, y_{ext}, z_{ext}$, which is local for the region "*ext*". The region formed with the slot cavity in the metal end wall of thickness h, which is general for the sections "*in*" and "*ext*", is denoted by the index "*v*". The local coordinate system x_v, y_v, z_v is introduced in this region.

Let surfaces of waveguide section ends $(z_{in,ext} = 0)$ be generally characterized by different distributed impedances $\overline{Z}_s^{in,ext}$ of which complex values are considered to be constant. Here $\overline{Z}_s^{in,ext} = Z_s^{in,ext}/Z_0$ are the surface impedances normalized to the wave impedance of free space $Z_0 = 120\pi\ \Omega$.

Immediately we note that the solution of boundary–value problems using the Leontovich–Shchukin impedance boundary conditions [1.2, 6.1], which will be used here, is an approximate solution in nature. More precisely this solution is the first term of asymptotic expansion of the exact solution of the problem in power of the small parameter $|\overline{Z}_s|$ ($|\overline{Z}_s| << 1$). Therefore it should be specified that it is supposed that $|\overline{Z}_s^{in,ext}| << 1$ in the considered case. In this case the boundary conditions become extraneous, i.e., a value of the surface impedance does not depend on the pattern of an exciting field. Also it should be kept in mind that the impedance conditions are inapplicable near the edges of verges [6.2]. Really, the thickness of the skin-layer Δ^0 must be small both in comparison with the dimensions of the body in all directions and in comparison with curvature radii of its surface [6.1]. In this connection it has turned out to be impossible to require the fulfillment of impedance boundary conditions at the ends of waveguide sections in the neighborhood of edges of the slot of the width $(10 \div 100)\Delta^0$. Formally for the solution of a boundary–value problem, it is necessary to exclude these boundary regions (the edge strip) from the total area of the impedance surface of the end. Using this approach, we maintain accuracy of the approximate solution of the boundary–value problem, and we can use the known methods developed for slots with perfectly conductive verges for studying fields of a slot aperture.

We note the continuity conditions of tangential components of a total magnetic field at the apertures of a coupling slot S^{in} ($z_{in} = 0$, $z_v = 0$) and S^{ext} ($z_{ext} = 0$, $z_v = -h$):

$$
\begin{cases}
\vec{H}^{in}_{0\tau}\{\vec{r}_{in}\} + \vec{H}^{in}_{\tau}\{\vec{r}_{in},\, \vec{E}^{in}_{sl}\} = \vec{H}^{v}_{\tau}\{\vec{r}_{v}\,,\, \vec{E}^{in}_{sl}\} + \vec{H}^{v}_{\tau}\{\vec{r}_{v}\,,\, \vec{E}^{ext}_{sl}\} \Big|_{\text{on } S^{in}}\,, \\[2mm]
\vec{H}^{v}_{\tau}\{\vec{r}_{v}\,,\, \vec{E}^{in}_{sl}\} + \vec{H}^{v}_{\tau}\{\vec{r}_{v}\,,\, \vec{E}^{ext}_{sl}\} = \vec{H}^{ext}_{\tau}\{\vec{r}_{ext}\,,\, \vec{E}^{ext}_{sl}\} \Big|_{\text{on } S^{ext}}\,,
\end{cases}
\tag{6.1}
$$

where $\vec{E}^{in}_{sl}$ and $\vec{E}^{ext}_{sl}$ are the electric fields at the corresponding apertures of the slot, $\vec{H}^{in}$, $\vec{H}^{v}$ and $\vec{H}^{ext}$ are the magnetic fields excited by the fields $\vec{E}^{in,ext}_{sl}$ in the space regions under consideration.

It is known [2.1] that $|E^{in,ext}_{sl\xi}| >> |E^{in,ext}_{sls}|$ (where s and ξ are the local coordinates related to the slot in the way shown in Fig. 6.1b) is universally valid for narrow slots except for the small regions of $\{d/\lambda\}$ approximately on their ends. Therefore we suppose that the vectors $\vec{E}^{in,ext}_{sl}$ in the set of equations (6.1) have only ξ-components, and the magnetic fields have only s-components. Then according to [3.1] the dependences of fields $E^{in,ext}_{sl\xi}$ on coordinates $s_{in,ext}$ and $\xi_{in,ext}$ can be assumed to be equal to

$$
E^{in,ext}_{sl\xi} = E^{in,ext}_{0}\, f_{in,ext}(s_{in,ext})\, \chi_{in,ext}(\xi_{in,ext}),
\tag{6.2}
$$
$$
f_{in,ext}(-L) = f_{in,ext}(+L) = 0,
$$

where $E^{in,ext}_{0}$ are the amplitudes of the fields, $f_{in,ext}(s_{in,ext})$ are the functional dependences on longitudinal coordinates at the slot apertures, $\chi_{in,ext}(\xi_{in,ext})$ are the functions taking into account physically justified behavior of an electric field in verges of a slot cavity and satisfying the normalization requirement

$$
\int_{-d/2}^{d/2} \chi_{in,ext}(\xi_{in,ext})\, d\xi_{in,ext} = 1.
\tag{6.3}
$$

Using the results of study [3.1], in the considered boundary–value problem it is permissible that $E^{in}_{0}\, f_{in}(s_{in}) \approx E^{ext}_{0}\, f_{ext}(s_{ext}) = E_{0}f(s) = E_{sl\xi}(s)$ to an accuracy of the terms of the order of $\{(hd)/\lambda^{2}\}$. Taking into account the normalization condition (6.3) after substitution of (6.2) in (6.1) the equation set can be reduced to one equation with "equivalent" slot width d_{e}:

$$
H^{in}_{0s}\{s, \xi\} + H^{in}_{s}\{s, \xi;\, E_{sl\xi}\} = H^{ext}_{s}\{s, \xi;\, E_{sl\xi}\} \Big|_{z_{in,ext}}.
\tag{6.4}
$$

In this case the region of the slot cavity v is eliminated from consideration, and in equation (6.4) d_e is related to real slot width d and wall thickness h by the approximate relationships (3.15) and (3.16).

To change the unknown electric field in the slot E_{sl_ξ} to equivalent currents distributed on its surface in equation (6.4) it is formally necessary to close the slot aperture in such a way that homogeneity of the surface in which the slot is made should not be violated from the side of the considered region. So, to obtain a perfectly conductive plane it is necessary to metallize the slot aperture. In the considered case of impedance surfaces according to the Leontovich–Shchukin boundary conditions $[\vec{n}, \vec{E}^{in,ext}] = -\overline{Z}_s^{in,ext}[\vec{n}, [\vec{n}, \vec{H}^{in,ext}]]$ ($\vec{n}$ is the normal to the surface, this normal is directed into the interior of the impedance body) it is necessary to take also into account "one-way" electric currents $\vec{J}_{in,ext}^E = \frac{1}{Z_0}[\vec{n}, \vec{H}_{sl}^{in,ext}]$ at that $\vec{J}_s^M = -\overline{Z}_s^{in,ext}[\vec{n}, \vec{J}_{in,ext}^E]$ besides for the magnetic current $\vec{J}^M$ at the slot aperture. We emphasize that in each of the connected waveguide sections the equivalent electric current will have a different value because it is determined through values of the surface impedances $\overline{Z}_s^{in,ext}$, which are generally assumed to be different. It must be said that "maintenance" of homogeneity of the boundary surfaces and introduction of the functionally related surface equivalent currents $\vec{J}^M$ and $\vec{J}_{in,ext}^E$ (where only $\vec{J}^M$ is unknown) does not contradict the requirements of assurance of uniqueness of the boundary-value problem solution. The "pair" assignment of equivalent currents on the impedance surface allows one to represent the electromagnetic field in semi-infinite waveguide sections simultaneously as superposition of waves of both magnetic and electric kinds using the corresponding components of the tensor Green's functions for the Hertz's vector potentials [1.3, 6.3].

Thus, equation (6.4) can be written in the following form:

$$H_{0s}^{in}\{s, \xi\} + H_s^{in}\{s, \xi; J_s^M\} + H_s^{in}\{s, \xi; J_{\xi in}^E\} = H_s^{ext}\{s, \xi; J_s^M\} + H_s^{ext}\{s, \xi; J_{\xi ext}^E\},$$
(6.5)

where $J_{\xi in,ext}^E = \frac{1}{\overline{Z}_s^{in,ext}} J_s^M$.

At the solution of equation (6.5) one can assume that $|\xi - \xi'| \cong \frac{d_e}{4}$ (quasi-one-dimensional approximation) in the dependences of field components on transverse coordinates with sufficient degree of accuracy, as it is accepted in the theory of thin dipole antennas and can be rightfully used for narrow slot radiators [2.12]. Then after simple transformations we finally obtain the integral–differential equation in the unknown magnetic slot current $J_s^M = J(s)$ in the following form:

$$\left(\frac{d^2}{ds^2} + k^2\right) \int_{-L}^{L} J(s')[G_s^{in}(s, s') + G_s^{ext}(s, s')]ds' +$$

$$+ ik \int_{-L}^{L} J(s')[\overline{Z}_s^{in} \tilde{G}_s^{in}(s, s') + \overline{Z}_s^{ext} \tilde{G}_s^{ext}(s, s')]ds' = -i\omega H_{0s}^{in}(s).$$
(6.6)

Here, $H_{0s}^{in}(s)$ is the projection of the field of impressed sources onto the slot axis, $G_s^{in,ext}(s, s')$ are the s-components of the quasi-one-dimensional tensor magnetic Green's functions $\hat{G}^M(\vec{r}_{in,ext}, \vec{r}'_{in,ext})$ for vector potential of corresponding volumes (see Appendix A),

$$\tilde{G}_s^{in,ext}(s, s') = \frac{\partial}{\partial z} G_s^{in,ext}\left(\begin{array}{c} x_{in,ext}(s),\, y_{0in,ext},\, z_{in,ext}; \\ x'_{in,ext}(s'),\, y_{0in,ext} + \dfrac{d_e}{4},\, 0 \end{array} \right)$$

at substitution of $z_{in,ext} = 0$ after derivation. It is necessary to note that at derivation of equation (6.6) it was taken into account that (in the case when the impressed sources are situated on the impedance surface of the end of a semi-infinite rect-angular waveguide) the components of the tensor of the electric Green's function $\hat{G}^E(\vec{r}_{in,ext}, \vec{r}'_{in,ext})$ are related to the components $\hat{G}^M(\vec{r}_{in,ext}, \vec{r}'_{in,ext})$ in the follow-ing way [1.3, 6.3]

$$G_x^E(\vec{r}_{in,ext}, \vec{r}'_{in,ext}) = (\overline{Z}_s^{in,ext})^2 G_y^M(\vec{r}_{in,ext}, \vec{r}'_{in,ext})$$

and

$$G_y^E(\vec{r}_{in,ext}, \vec{r}'_{in,ext}) = (\overline{Z}_s^{in,ext})^2 G_x^M(\vec{r}_{in,ext}, \vec{r}'_{in,ext}).$$

In the case when $\overline{Z}_s^{in,ext} = 0$ equation (6.6) is transformed to the equation (5.9) concerning the magnetic current in a slot coupling two semi-infinite rectangular waveguides.

6.2 Solution of the Equation for a Magnetic Current

We will apply the induced magnetomotive forces method (IMMFM) (see Chap. 4) for the approximate analytical solution of equation (6.6). As approximation for the magnetic current we will use distribution (5.19), which has been obtained due to solution (by means of the averaging method) of the problem of diffraction of a fundamental mode by a narrow symmetrical slot that is made in the resonant iris of an infinite rectangular waveguide (see Chap. 5):

$$J(s) = J_0 f(s) = J_0 \left(\cos ks \cos \frac{\pi}{a_{in}} L - \cos kL \cos \frac{\pi}{a_{in}} s \right), \tag{6.7}$$

where J_0 is the unknown complex amplitude.

Approximation (6.7) obeys boundary conditions (6.2) and is natural in the con-sidered case of symmetrical arrangement of the slot with respect to the axial line of the broad wall of the waveguide section "*in*", where impressed sources of the electromagnetic field are placed.

Substituting (6.7) in equation (6.6) and carrying out necessary transformations according to Chap. 4, we obtain the solution for the sought current in the form

$$J(s) = -\left(\frac{i\omega}{2k}\right)\frac{\cos ks \cos \dfrac{\pi}{a_{in}}L - \cos kL \cos \dfrac{\pi}{a_{in}}s}{Y^{in}(kL,\overline{Z}_s^{in}) + Y^{ext}(kL,\overline{Z}_s^{ext})} \times \int_{-L}^{L} f(s)H_{0s}^{in}(s)\,ds, \quad (6.8)$$

where $Y^{in,ext}$ are the corresponding admittances of the slot in the volumes "in" and "ext" determined by means of the expressions

$$Y^{in,ext}(kL,\overline{Z}_s^{in,ext}) = \frac{1}{2k}\int_{-L}^{L} f(s)\left[\begin{array}{c}\left(\dfrac{d^2}{ds^2}+k^2\right)\displaystyle\int_{-L}^{L} f(s')G_s^{in,ext}(s,s')ds'+\\[2mm] +ik\overline{Z}_s^{in,ext}\displaystyle\int_{-L}^{L} f(s')\tilde{G}_s^{in,ext}(s,s')ds'\end{array}\right]ds. \quad (6.9)$$

Using the expression (A.4) for the Green's function $G_s^{in,ext}(s,s')$ from Appendix A, we find explicit admittances of the slot in each of the coupled waveguide sections from relations (6.9) (see Appendix C).

Taking into account that in the case under consideration the slot is excited by the fundamental mode $H_{10}(x_{in},z_{in}) = H_0 \sin\frac{\pi x_{in}}{a_{in}}e^{-i\gamma z_{in}}$, where H_0 is its amplitude and $\gamma = \sqrt{k^2-(\pi/a_{in})^2}$ is the propagation constant in the waveguide "in," it is valid $H_{0s}^{in}(s) = 2H_0\cos\frac{\pi s}{a_{in}}$ in expression (6.8). After integration in (6.8) we obtain the final formula for determination of the magnetic current in the slot aperture

$$J(s) = -\frac{i\omega}{k^2}H_0\frac{g(kL)}{Y^{in}(kL,\overline{Z}_s^{in}) + Y^{ext}(kL,\overline{Z}_s^{ext})}\left(\cos ks \cos\frac{\pi}{a_{in}}L - \cos kL \cos\frac{\pi}{a_{in}}s\right), \quad (6.10)$$

where

$$g(kL) = 2\cos\frac{\pi L}{a_{in}}\frac{\sin kL \cos\dfrac{\pi L}{a_{in}} - \left(\dfrac{\pi}{ka_{in}}\right)\cos kL \sin\dfrac{\pi L}{a_{in}}}{1-[\pi/(ka_{in})]^2}$$

$$-\cos kL\frac{\sin\dfrac{2\pi L}{a_{in}}+\dfrac{2\pi L}{a_{in}}}{[2\pi/(ka_{in})]}.$$

Thus, we have realized the approximate analytical solution (6.10) of integral–differential equation (6.6) for the magnetic current in the slot aperture. This enables us to determine coefficients of the scattering matrix of stepped rectangular waveguides coupling with an impedance slot iris.

The expression for the power reflection coefficient $|S_{11}|^2$ at slot obstacle in a semi-infinite rectangular waveguide "*in*" with impedance end at single-mode operating has the following form:

$$|S_{11}|^2 = \left| \frac{1 - (\gamma/k)\overline{Z}_s^{in}}{1 + (\gamma/k)\overline{Z}_s^{in}} - \frac{8\pi\gamma\, g^2(kL)}{ia_{in}b_{in}k^3[Y^{in}(kL,\overline{Z}_s^{in}) + Y^{ext}(kL,\overline{Z}_s^{ext})]} \right.$$

$$\left. \times \frac{1 + (\overline{Z}_s^{in})^2}{1 + (\gamma/k)\overline{Z}_s^{in}} \right|^2 . \tag{6.11}$$

The second item in formula (6.11) is determined by the field, which is excited in a waveguide section by proper magnetic current (6.10) of the slot. This item is found by means of the magnetic Green's function (A.4). The first addend is the coefficient of H_{10}-wave reflection at impedance end without a slot. This item is determined by means of Green's magnetic function (A.5) for a longitudinal current that excites the semi-infinite rectangular waveguide with impedance end assuming arrangement of a point source at $z' \to \infty$. Note that in the case of perfectly conductive end ($\overline{Z}_s^{in} = 0$) the first item becomes equal to unity (as it must be).

The number of propagated modes M in the waveguide section "*ext*", certainly, will depend on the dimensions of its cross section $\{a_{ext} \times b_{ext}\}$. In keeping with the above-mentioned, the scattering matrix of a waveguide coupling will contain a definite number of elements, which are power transmission coefficients of the propagated modes (or power transformation coefficients of the modes). They are determined as the ratio of the transmitted power P_{mn}^{tr} in the waveguide "*ext*" per each mode separately to the power P_{10}^{inc} of the incident H_{10}-wave in the waveguide "*in*", i.e.,

$$P_{mn} = \frac{P_{mn}^{tr}}{P_{10}^{inc}}, \tag{6.12}$$

where $P_{10}^{inc} = H_0^2 \left(\frac{a_{in}b_{in}k}{2\gamma} \right)$.

In expression (6.12) the quantity P_{mn}^{tr} is the power transferred in the waveguide section "*ext*" by the mode with the indexes (m, n). It is determined by means of the found equivalent currents in the slot aperture and the dyadic Green's functions. Since the particular case of H-plane coupling of waveguides is considered as an example below in the study, here we will write down the explicit expressions for P_{m0} of waveguide modes of a magnetic type:

$$P_{m0} = \frac{a_{ext}b_{ext}\gamma}{a_{in}b_{in}\gamma_{m0}} \left| \frac{8\pi\gamma_{m0} \sin \dfrac{\pi x_{0ext}}{a_{ext}} g(kL) f_{m0}(kL)}{a_{ext}b_{ext}k^3} \right.$$

$$\left. \times \frac{F(\gamma_{m0}, \overline{Z}_s^{ext})}{[Y^{in}(kL,\overline{Z}_s^{in}) + Y^{ext}(kL,\overline{Z}_s^{ext})]} \right|^2 , \tag{6.13}$$

where

$$\gamma_{m0} = \sqrt{k^2 - (m\pi/a_{ext})^2}, \quad F(\gamma_{m0}, \overline{Z}_s^{ext}) = \frac{1 + (\overline{Z}_s^{ext})^2}{1 + (\gamma_{m0}/k)\overline{Z}_s^{ext}},$$

$$f_{m0}(kL) = 2\cos\frac{\pi L}{a_{in}} \frac{\sin kL \cos\dfrac{m\pi L}{a_{ext}} - \left(\dfrac{m\pi}{ka_{ext}}\right)\cos kL \sin\dfrac{m\pi L}{a_{ext}}}{1 - [m\pi/(ka_{ext})]^2}$$

$$- 2\cos kL \frac{\left(\dfrac{\pi}{ka_{in}}\right)\sin\dfrac{\pi L}{a_{in}}\cos\dfrac{m\pi L}{a_{ext}} - \left(\dfrac{m\pi}{ka_{ext}}\right)\cos\dfrac{\pi L}{a_{in}}\sin\dfrac{m\pi L}{a_{ext}}}{[\pi/(ka_{in})]^2 - [m\pi/(ka_{ext})]^2}.$$

In the case under consideration the complete scattering matrix of waveguide coupling, besides the coefficients $|S_{11}|^2$ and P_{m0}, will also contain the quantity P_σ. This is the loss power in impedance coatings of ends of both coupled waveguide sections. The loss power can be determined from the condition of energy balance fulfillment:

$$|S_{11}|^2 + \sum_{m=1}^{M} P_{m0} + P_\sigma = 1. \tag{6.14}$$

In the case of absence of loss ($P_\sigma = 0$) it is necessary to use condition (6.14) for validation of algorithms of mathematical simulation. The expressions for the $Y^{in,ext}(kL, \overline{Z}_s^{in,ext})$ admittances are given in Appendix C.

6.3 Surface Impedance of the Coating with the ε_1 and μ_1 Homogeneous Parameters

For some examples of physical realization of impedance surfaces, we give the formulas determining the quantity $\overline{Z}_s$ since they are necessary for numerical calculations. For this purpose, we consider the model problem of a plane electromagnetic wave normally incident on a dielectric layer of thickness h_d with complex permittivity ε_1 and permeability μ_1, and wave number $k_1 = k\sqrt{\varepsilon_1\mu_1}$. The layer separates two half-spaces. The upper half-space, from which the plane wave is incident, is free space ($\varepsilon = \mu = 1$), while the second half-space is described by the material parameters ε_2, μ_2.

Making use of the boundary conditions for the electric and magnetic field components on both surfaces of a dielectric layer, it is easy to find the solution of the boundary-value problem. Comparing this solution with the requirements that the Leontovich–Shchukin impedance boundary condition on the upper boundary of the dielectric layer must be fulfilled, we obtain a rigorous expression for the distributed surface impedance in the form:

$$\overline{Z}_s = \overline{Z}_1 \frac{i\overline{Z}_1 \text{tg}\,(k_1 h_d) + \overline{Z}_2}{\overline{Z}_1 + i\overline{Z}_2 \text{tg}\,(k_1 h_d)}, \tag{6.15}$$

where $\overline{Z}_1 = \sqrt{\mu_1/\varepsilon_1}$ and $\overline{Z}_2 = \sqrt{\mu_2/\varepsilon_2}$.

For a magnetodielectric layer of the thickness h_d on a metallic surface on substituting $\overline{Z}_2 = 0$ we obtain the formula, which is used to calculate the value $\overline{Z}_s$ in [6.4]:

$$\overline{Z}_s = i\sqrt{\frac{\mu_1}{\varepsilon_1}} \text{tg}\,(\sqrt{\varepsilon_1 \mu_1} k h_d). \tag{6.16}$$

We note that in this case the arbitrary incident field formula (6.16) is approximate, and it becomes more accurate the better the inequality $|\varepsilon_1 \mu_1| >> 1$ is fulfilled. So, for example, if the incident field is the wave of the H_{10} mode, propagating in a rectangular waveguide, then formulas (6.15) and (6.16) have the form:

$$\overline{Z}_s = \overline{Z}_1 \frac{i\overline{Z}_1 \text{tg}\,(\gamma_1 h_d) + \overline{Z}_2}{\overline{Z}_1 + i\overline{Z}_2 \text{tg}\,(\gamma_1 h_d)}, \tag{6.17}$$

$$\overline{Z}_s = i\frac{k_1}{\gamma_1}\sqrt{\frac{\mu_1}{\varepsilon_1}} \text{tg}\,(\gamma_1 h_d), \quad \gamma_1 = \sqrt{k_1^2 - k_c^2}, \tag{6.18}$$

and when condition $|\varepsilon_1 \mu_1| >> 1$ is performed, they transit into expressions (6.15) and (6.16), correspondingly.

For the electrically thin layer ($|k_1 h_d| << 1$, the quasi-stationary approximation [6.2]) from (6.16) and (6.18), it follows that $\overline{Z}_s \approx i k \mu_1 h_d$ [6.2], that is, the normalized surface impedance does not depend on permittivity of material.

Supposing that in the expression (6.15) $\overline{Z}_2 = \overline{Z}_1$ under the condition $|\varepsilon_1| >> 1$ and $\mu_1 = 1$, taking into account that $\varepsilon_1 = \text{Re}\varepsilon_1 + \frac{4\pi\sigma_1}{i\omega}$, where σ_1 is the material conductivity, we can obtain a known formula to take into account the skin-effect of the conductor:

$$\overline{Z}_s = \frac{1+i}{Z_0 \sigma_1 \Delta^0}, \tag{6.19}$$

where $\Delta^0 = \frac{1}{30\sqrt{2\pi\sigma_1\omega}}$ is the depth of penetration of the electromagnetic field into a conductor.

In the case of a thin conductive film with the thickness h_R ($(h_R/\Delta^0) << 1$), which is covered on a magnetodielectric layer located on a metallic plane, the surface impedance, according to (6.15), is given by:

$$\overline{Z}_{sR} = \frac{\overline{R}_{sR}}{1 + \overline{R}_{sR}/\overline{Z}_s}, \quad \overline{R}_{sR} = \frac{1}{Z_0 \sigma_1 h_R}, \tag{6.20}$$

where $\overline{Z}_s$ is defined by the formulas (6.16) and (6.18).

Further we will use representations (6.16) and (6.20) for $\overline{Z}_s$. However, if the analysis of other structures of impedance coverings is required, the expressions for $\overline{Z}_s$ can be obtained by different methods, as it is demonstrated in the following section.

6.4 Surface Impedance of the Magnetodielectric Layer with Inhomogeneous Permittivity on the Perfectly Conductive Surface

Modern technology of thin-film coverings allow us to obtain both homogeneous (in the direction perpendicular to the completely conducting plane of the basis) and inhomogeneous structures [6.5]. In this section we have obtained the approximate analytical expressions for distributed surface impedance of the magnetodielectric layer with inhomogeneous permittivity, which is located on the infinite conductive plane, and they are just at rather small change of the value of permittivity within the considered layer.

Let a plane monochromatic electromagnetic wave, for which the component of the electric field equals to $E_x(z) = E_{0x}e^{-ikz}$ (E_{0x} is the amplitude) incident on the magnetodielectric layer with the thickness $2h_d$ ($-\infty < x, y < \infty; -h_d \leq z \leq h_d$, where $\{x, y, z\}$ is Cartesian coordinate system) and with permeability μ_1 and permittivity ε_1, located on the infinite conductive plane at $z = h_d$, perpendicular to the surface layer from the region $z = -\infty$ (half-space). Then for this layer the distributed surface impedance, normalized on a wave impedance of free space $Z_0 = 120\pi$ Ω, will be defined by expression [6.2]:

$$\overline{Z}_s = \frac{E_{0x}(-h_d)}{H_{0y}(-h_d)}. \tag{6.21}$$

To obtain $E_{0x}(-h_d)$ and $H_{0y}(-h_d)$ it is necessary to know the fields $E_x(z)$ and $H_y(z)$ inside the magnetodielectric layer, where they satisfy the following differential equations at $\mu_1 = const$ and $\varepsilon_1 = \varepsilon_1(z)$:

$$\frac{d^2 E_x(z)}{dz^2} + k^2 \mu_1 \varepsilon_1(z) E_x(z) = 0, \tag{6.22a}$$

$$H_y(z) = \frac{i}{k\mu_1} \frac{dE_x(z)}{dz}. \tag{6.22b}$$

The solution of the equations (6.22) together with boundary conditions on the surfaces of the layer at $z = \pm h_d$ gives opportunity to define the values $E_{0x}(-h_d)$, $H_{0y}(-h_d)$ and the value defined of the surface impedance $\overline{Z}_s$, correspondingly. We note that equations (6.22) are just at the arbitrary functional dependence $\varepsilon_1(z)$, and ratio (6.21) is rigorous at a normal incident of a plane wave on a flat boundary of magnetodielectric.

For a limited number of laws of change $\varepsilon_1(z)$ we can obtain exact solutions of the equations (6.22) [6.6]. However, they are very complicated and in every concrete case are obtained through a definite class of special functions. At relatively small changes of the permittivity $\varepsilon_1(z)$ values within the layer we can obtain approximate solutions in the class of elementary functions, which have sufficient accuracy and simplicity. Let us consider some of them, which will allow us to obtain expressions for the surface impedance $\overline{Z}_s$ in an analytical form.

6.4.1 Power Law of the $\varepsilon_1(z)$ Change

In this case it is possible to represent $\varepsilon_1(z)$ in the following form:

$$\varepsilon_1(z) = \varepsilon_1(0)[1 - \varepsilon_r f(z)], \tag{6.23}$$

where $\varepsilon_r = \frac{\varepsilon_1(0) - \varepsilon(-h_d)}{\varepsilon_1(0)}$ is the value of relative change of permittivity within the layer ($|\varepsilon_r| << 1$) and $f(z) = \left(-\frac{z}{h_d}\right)^n$, ($n = 1, 2, 3...$) is the fixed function. Then the equation (6.22a) transforms into the non-uniform differential equation with constant coefficients:

$$\frac{d^2 E_x(z)}{dz^2} + k_1^2 E_x(z) = \varepsilon_r k_1^2 f(z) E_x(z), \tag{6.24}$$

where $k_1^2 = k^2 \mu_1 \varepsilon_1(0)$.

Further, due to the method of variation of arbitrary constants we have, considering the right part of equation (6.24) fixed:

$$E_x(z) = C_1 e^{-ik_1 z} + C_2 e^{ik_1 z} + \varepsilon_r k_1 \int\limits_{-h_d}^{z} f(z') E_x(z') \sin k_1(z - z') dz'. \tag{6.25}$$

We will search the solution of the integral equation (6.25) in the form of expansion $E_x(z)$ on the ε_r small parameter:

$$E_x(z) = E_{x0}(z) + \varepsilon_r E_{x1}(z) + \varepsilon_r^2 E_{x2}(z) + ... + \varepsilon_r^m E_{xm}(z) + ... \tag{6.26}$$

Then, substituting (6.26) into equation (6.25) and being limited by components of the null and first order of smallness, we obtain the following expressions for the electric and magnetic fields inside the layer:

$$
\begin{aligned}
E_x(z) &= C_1 e^{-ik_1 z}[1 + \varepsilon_r f_E(k_1 z)] + C_2 e^{ik_1 z}[1 + \varepsilon_r f_E^*(k_1 z)], \\
H_y(z) &= \frac{1}{Z_1} \left\{ C_1 e^{-ik_1 z}[1 + f_H(k_1 z)] - C_2 e^{ik_1 z}[1 + f_H^*(k_1 z)] \right\}.
\end{aligned}
\tag{6.27}
$$

Here "$*$" is the sign of complex conjugation, $\overline{Z}_1 = \sqrt{\mu_1/\varepsilon_1(0)}$, and the functions $f_E(k_1z)$ and $f_H(k_1z)$ are defined by the law of change $\varepsilon_1(z)$. Particularly, they are equal for the linear and square laws, correspondingly:

a) the linear law: $\varepsilon_1(z) = \varepsilon_1(0)\left(1 + \varepsilon_r\frac{z}{h_d}\right)$;

$$f_E(k_1z) = -\frac{1}{4k_1h_d}[(k_1z) + i(k_1z)^2 - i/2],$$

$$f_H(k_1z) = -\frac{1}{4k_1h_d}[-(k_1z) + i(k_1z)^2 + i/2].$$

(6.28)

b) the square law: $\varepsilon_1(z) = \varepsilon_1(0)\left(1 - \varepsilon_r\frac{z^2}{h_d^2}\right)$;

$$f_E(k_1z) = \frac{1}{(2k_1h_d)^2}\left[(k_1z)^2 - \frac{1}{2} + i\frac{2(k_1z)^3}{3} - i(k_1z)\right],$$

$$f_H(k_1z) = \frac{1}{(2k_1h_d)^2}\left[-(k_1z)^2 + \frac{1}{2} + i\frac{2(k_1z)^3}{3} + i(k_1z)\right].$$

(6.29)

We obtain searched expressions for the surface impedance defining unknown constants C_1 and C_2 from the boundary conditions of the components continuity at $z = -h_d$ and the null equality of the electrical component at $z = h_d$ in (6.27):

$$\overline{Z}_s = \overline{Z}_1 \frac{\begin{cases} [1 + \varepsilon_r f_E(-k_1h_d)][1 + \varepsilon_r f_E^*(k_1h_d)]e^{i4k_1h_d} \\ -[1 + \varepsilon_r f_E^*(-k_1h_d)][1 + \varepsilon_r f_E(k_1h_d)] \end{cases}}{\begin{cases} [1 + \varepsilon_r f_H^*(-k_1h_d)][1 + \varepsilon_r f_E(k_1h_d)] \\ +[1 + \varepsilon_r f_H(-k_1h_d)][1 + \varepsilon_r f_E^*(k_1h_d)]e^{i4k_1h_d} \end{cases}}.$$

(6.30)

At $\varepsilon_r = 0$ and the change $2h_d \rightarrow h_d$, formula (6.30) transits into ratio (6.16) obtained before, for impedance of the homogeneous magnetodielectric layer of the thickness h_d on the perfectly conductive surface.

Neglecting the components of the ε_r^2-order in the formula (6.30) and making change $2h_d \rightarrow h_d$, finally, we have:

a) for the linear law of change $\varepsilon_1(z)$:

$$\overline{Z}_s = i\overline{Z}_1\frac{tg(k_1h_d)}{1 + \varepsilon_r f_{Lin}(k_1h_d)tg(k_1h_d)},$$

$$f_{Lin}(k_1h_d) = \left(\frac{1}{2k_1h_d} + \frac{i}{2}\right).$$

(6.31)

b) for the square law of change $\varepsilon_1(z)$:

$$\overline{Z}_s = i\overline{Z}_1 \frac{\text{tg}(k_1 h_d) + \varepsilon_r f_{Sq1}(k_1 h_d)}{1 + \varepsilon_r f_{Sq2}(k_1 h_d)\text{tg}(k_1 h_d)},$$

$$f_{Sq1}(k_1 h_d) = \frac{1}{k_1 h_d} - \frac{\text{tg}(k_1 h_d)}{(k_1 h_d)^2}, \quad f_{Sq2}(k_1 h_d) = \frac{k_1 h_d}{6}. \tag{6.32}$$

One must note, that the obtained approximate solution (6.27) of the equation (6.24) (and the expression for impedance (6.30), correspondingly), is not just for the power law of change $\varepsilon_1(z)$ because representation of permittivity in form (6.23) is possible for other functions $f(z)$, e.g., trigonometric ones: $f(z) = \left(1 - \cos \frac{\pi z}{2h_d}\right)$, $f(z) = -\sin \frac{\pi z}{2h_d}$.

6.4.2 Exponential Law of the $\varepsilon_1(z)$ Change

Let us consider that permittivity of the magnetodielectric layer of the h_d $(0 \le z \le h_d)$ thickness varies due to the exponential law

$$\varepsilon_1(z) = \varepsilon_1(0) e^{kz}. \tag{6.33}$$

In this case transformation (6.23), made in order to separate the ε_r small parameter, does not take place, though some other representations $\varepsilon_1(z)$ are possible, which allow us to reduce the initial equation (6.22a) to the integral equation with a small parameter and to apply the method, described above, for its solution. To our minds, at this functional dependence $\varepsilon_1(z)$ it is more convenient to use the approximate solution of equation (6.22a) in a form of the so-called WKB-approach [1.8, 6.6]:

$$E_x(z) = \frac{1}{\sqrt[4]{\varepsilon_1(z)}} \left\{ C_1 e^{-ik\sqrt{\mu_1} \int_0^z \sqrt{\varepsilon_1(z)}dz} + C_2 e^{ik\sqrt{\mu_1} \int_0^z \sqrt{\varepsilon_1(z)}dz} \right\}. \tag{6.34}$$

Let us note that expression (6.34) is just if the $\varepsilon_1(z)$ function does not have nulls or poles on the $[0, z]$ interval. The criteria to apply the WKB-approach is the following inequality [1.8]:

$$\left| \frac{d\varepsilon_1(z)/dz}{k\sqrt{\mu_1}\,[\varepsilon_1(z)]^{3/2}} \right|_{\max} << 1, \tag{6.35}$$

which has form $\left| \frac{1}{\sqrt{\mu_1 \varepsilon_1(0)}} \right| << 1$ in our case.

Substituting (6.33) into (6.34) and imposing restriction $(h_d/\lambda)^2 << 1$ after transformations we get:

$$E_x(z) = \frac{e^{-kz/4}}{\sqrt[4]{\varepsilon_1(0)}} \left(C_1 e^{-ik_1 z} + C_2 e^{ik_1 z} \right),$$

$$H_y(z) = \frac{1}{Z_1} \frac{e^{-kz/4}}{\sqrt[4]{\varepsilon_1(0)}} \left\{ C_1 e^{-ik_1 z} f_H[\varepsilon_1(0)\mu_1] - C_2 e^{ik_1 z} f_H^*[\varepsilon_1(0)\mu_1] \right\},$$

(6.36)

where $f_H[\varepsilon_1(0)\mu_1] = 1 - \frac{i}{4\sqrt{\varepsilon_1(0)\mu_1}}$. Due to the requirements of boundary conditions performance at $z = 0$ and $z = h_d$ we define the arbitrary constants C_1 and C_2 and later on the expression for the normalized surface impedance at the exponential law of change of permittivity inside the magnetodielectric layer:

$$\overline{Z}_s = i\overline{Z_1} \frac{tg(k_1 h_d)}{1 + f_{Exp}[\varepsilon_1(0)\mu_1] tg(k_1 h_d)},$$

$$f_{Exp}[\varepsilon_1(0)\mu_1] = \frac{1}{4\sqrt{\varepsilon_1(0)\mu_1}},$$

(6.37)

transitting to the already known ratio $\overline{Z}_s \approx ik\mu_1 h_d$ at $|k_1 h_d| << 1$.

We must emphasize the fact that formulas (6.31), (6.32) and (6.37) for the distributed surface impedance of the magnetodielectric layer with inhomogeneous permittivity, located on the infinite conductive plane, are just at different laws of change of permittivity inside the layer and obtained by different methods, have similar structure and are reduced to the same ratios in the restricted case $\varepsilon_1 = const$ and $|k_1 h_d| << 1$. Using these methods, the obtained results permit us to generalize them for the case where a metallic film of a thickness less than skin-layer thickness is covered on the magnetodielectric layer located on the metallic surface. Then the surface impedance of such a structure will be defined by expression (6.20), and $\overline{Z}_s$ is calculated due to formulas (6.31), (6.32), and (6.37) according to the law of distribution of permittivity of the layer.

6.4.3 Numerical Examples

We give some results of numerical calculations of the value of the distributed surface impedance, which illustrate the effect of that or other laws of permittivity change on the values of active and reactive parts of the impedance of the magnetodielectric layer located on the infinite metallic plane. Because of dependence of magnetodielectric parameters from the frequency, we limit ourselves by the corresponding single-mode standard rectangular waveguide of a centimeter band to define waves lengths.

Figure 6.2 represents the curves of real and imaginary parts of the complex surface impedance $\overline{Z}_s = \overline{R}_s + i\overline{X}_s$, depending on the thickness of the magnetodielectric layer made from the TDK IR-E110 material, for which, due to [6.4] $\varepsilon_1 = 8.84 - i0.084$ and $\mu_1 = 2.42 - 0.0825f - i0.994$, where f is the frequency [GHz] corresponding to the band of the H_{10}-mode of the rectangular waveguide by cross section 22.86×10.16 mm^2. As can be seen, a real impedance $\overline{R}_s$ has an

Fig. 6.2 The real and imaginary parts of surface impedance dependences from the layer thickness for a magnetodielectric layer on a metallic plane at $\lambda = 30.0$ mm

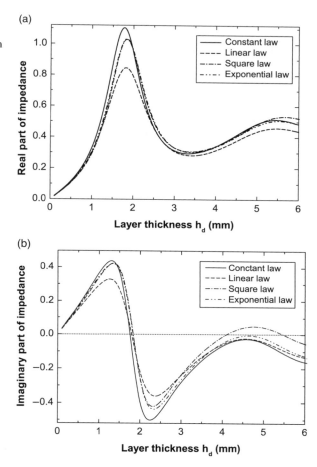

explicitly marked maximum for all considered laws of change $\varepsilon_1(z)$ (here and further $\varepsilon_r^2 = 0.04$ for the linear and square laws) at a definite layer thickness that is equal to a quarter of a wavelength in the magnetodielectric ($h_d \approx 1.8$ mm mm for $\lambda = 30$ mm), and it aspires to a constant value that equals $\mathrm{Re}\overline{Z}_1$ at losses presence when the layer thickness increases further. At the same time the dependence of the imaginary part $\overline{X}_s$ of the impedance from the layer thickness has alternating-sign character; moreover, for the fixed ε_1, μ_1 and λ at $h_d \approx 1.8$mm for all dependences $\varepsilon_1(z)$ they have $\overline{X}_s = 0$, and $\overline{R}_s = \overline{R}_{s_{max}}$, and the coefficient of plane wave reflection from such a structure will be minimal: the layer becomes the so-called "anti-reflecting coating". We note that the largest difference of the $\overline{Z}_s$ value from the case of a homogeneous magnetodielectric takes place at the linear law of distribution $\varepsilon_1(z)$.

Figure 6.3 gives the curves of the dependences of the surface impedance from a wavelength due to different laws of change of permittivity of a magnetodielectric

Fig. 6.3 The real and imaginary parts of surface impedance dependences from the wavelength for a magnetodielectric layer on a metallic plane at h_d=1.6 mm

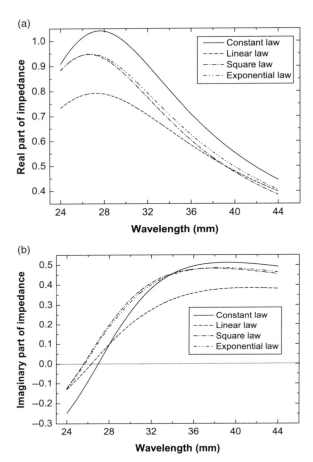

TDK IR-E110 and the layer thickness $h_d \approx 1.6$ mm, which corresponds to the value in [6.4], where the problem of radiation from a longitudinal slot in the broad wall of a rectangular waveguide ($a = 22.86$ mm; $b = 10.16$ mm) into half-space over the impedance plane has been solved. From the plots it follows that the inhomogeneous layer leads to the $\overline{R}_s$ decrease in a whole band of the H_{10}-mode of a waveguide; meanwhile, the imaginary part of impedance can be of both larger and smaller values, corresponding to the case of a homogeneous material. Let us note that the values of the surface impedance $\overline{Z}_s$ practically coincide for the square and exponential laws of change $\varepsilon_1(z)$ at the given layer thickness.

When solving the problem of an impedance slotted iris we used the Leontovich–Shchukin approximate impedance boundary condition, which assumes performance of the inequality $|\overline{Z}_s| << 1$. Such values of impedance take place for small thicknesses of the layer, as is shown, for example, in Fig. 6.4 (the same material, $h_d = 0.3$ mm). In this case both the real and imaginary parts of impedance

Fig. 6.4 The real and imaginary parts of surface impedance dependences from the wavelength for a magnetodielectric layer on a metallic plane at $h_d = 0.3$ mm

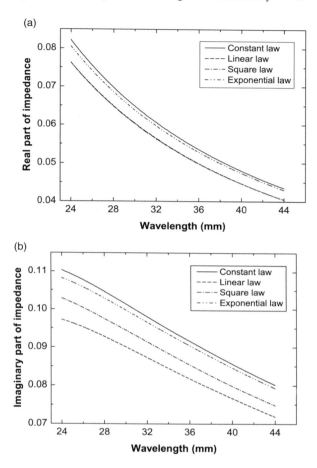

independent from the law of change of permittivity decrease monotonously at a wave length increase; moreover, the $\overline{X}_s$ value is always positive—that is, it has an inductive character.

At can be seen in Fig. 6.5 if a resistive film of a definite thickness with low conductivity (e.g., Nichrome) is covered on the magnetodielectric surface, then the real part $\mathrm{Re}\overline{Z}_{sR}$ of the surface impedance of such a structure has the same value and changes, practically, as in the previous case (Fig. 6.4a). The imaginary part of the $\mathrm{Im}\overline{Z}_{sR}$ impedance becomes sufficiently less, and it has maximum in a short wave part of waveguide single-mode band.

The analysis shows that the effect of inhomogeneity of magnetodielectric in the considered cases is relatively small—not more than 25% in comparison with a homogeneous layer—and it is sufficient for the linear law of distribution (in comparison with the square and exponential laws). The layer inhomogeneity characterized by one or another law of change of permittivity is an additional method to achieve

Fig. 6.5 The real and imaginary parts of surface impedance dependences from the wavelength for a magnetodielectric layer on a metallic plane covered by resistive film at: $\overline{R}_{sR} = 0.3$, $h_d = 0.3$ mm

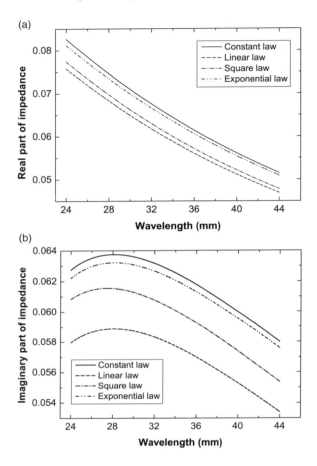

range extension of conducting of management of electrodynamic parameters of microwave wave-slotted devices.

6.5 Numerical Results

Based on the obtained problem solution the numerical analysis of coefficients of the scattering matrix of an impedance slot iris for a number of cases of realization of stepped joint of two rectangular waveguides are carried out. All calculations are carried out under the assumption that the excited region "*in*" is a hollow semi-infinite rectangular waveguide of standard cross section $a_{in} \times b_{in} = 23.0 \times 10.0$ mm^2 with thickness of end wall $h = 2.0$ mm and geometrical parameters of a slot: $d = 1.5$ mm, $x_{0in} = a_{in}/2$ and $y_{0in} = b_{in}/4$; the waveguide is used in single-mode operation. The other parameters of the diffraction problem are varied during calculations.

Fig. 6.6 The coefficients of the scattering matrix dependences from the wavelength for a slot impedance iris in a regular waveguide at:
$a_{in} = a_{ext} = 23.0$ mm,
$b_{in} = b_{ext} = 10.0$ mm,
$d = 1.5$ mm, $2L = 16.0$ mm,
$x_{0in} = a_{in}/2$, $y_{0in} = b_{in}/4$

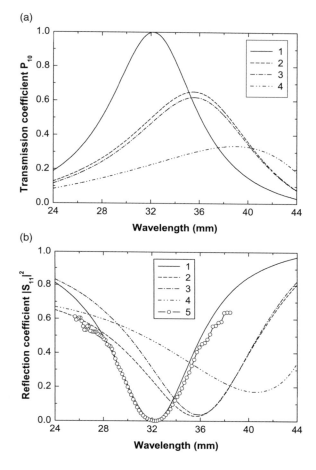

Figures 6.6 and 6.7 represent the range dependences of coefficients of the scattering matrix and loss power of a slot iris in a regular rectangular waveguide ($a_{ext} = a_{in} = 23.0$ mm, $b_{ext} = b_{in} = 10.0$ mm) for $2L = 16$ mm as a matter for convenience and completeness of the analysis of the numerical results. Curve 1 shows the results of calculating the characteristics at $\overline{Z}_s^{in,ext} = 0$. Certainly there is no curve 1 in Fig. 6.7 because the loss power is $P_\sigma = 0$ in the case of a perfectly conductive iris. As one can see from Fig. 6.6b the satisfactory agreement of theoretical (curve 1) and experimental (curve 5) results confirms both validity of the procedure and the accepted approximations and validity of construction of algorithms of mathematical simulation.

Figure 6.6 also represents the calculation dependences of energy characteristics of a slot iris over a working range of wavelengths of a rectangular waveguide at various values of complex impedances $\overline{Z}_s^{in,ext}$ with imaginary part of inductive type for a layer situated on metal of the magnetodielectric TDK IR-E110. In this case the

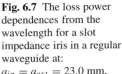

Fig. 6.7 The loss power dependences from the wavelength for a slot impedance iris in a regular waveguide at:
$a_{in} = a_{ext} = 23.0$ mm,
$b_{in} = b_{ext} = 10.0$ mm,
$d = 1.5$ mm, $2L = 16.0$ mm,
$x_{0in} = a_{in}/2$, $y_{0in} = b_{in}/4$

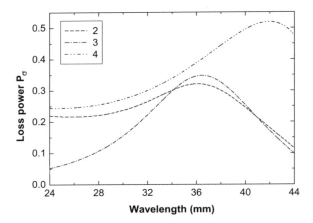

values $\overline{Z}_s^{in,ext}$ are determined by the formula (6.16). In Fig. 6.6 curve 2 corresponds to the case of coating of the iris surface from the side of the region "*in*" with the material TDK (in this case $\overline{Z}_s^{ext} = 0$), curve 3 corresponds to the case of coating of the iris surface from the side of the region "*ext*" with the material TDK ($\overline{Z}_s^{in} = 0$), curve 4 corresponds to the case of two-sided coating of the iris. Here it is assumed that the thickness of layers of the material TDK is $h_d = 0.2$ mm everywhere.

Naturally the two-sided impedance iris exerts more essential influence upon behavior of the energy dependences than one-sided ones. Also is observed nonequivalence of the cases of one-sided impedance coating of an iris ($\overline{Z}_s^{in} \rightarrow (6.16)$, $\overline{Z}_s^{ext} = 0$) and ($\overline{Z}_s^{in} = 0$, $\overline{Z}_s^{ext} \rightarrow (6.16)$). It follows from formula (6.11), which is asymmetrical with respect to the parameters $\overline{Z}_s^{in}$ and $\overline{Z}_s^{ext}$. Note that in the cases under consideration the displacement of the resonance wavelength of the iris λ_{res} (corresponds to minimum in dependences of power reflection coefficient $|S_{11}|^2$ in Fig. 6.6b) in the long-wave region of the working band of the waveguide is characterized by the considerable level of the loss power P_σ in impedance surfaces (Fig. 6.7). Thus for $\lambda_{res} = 41.0$ mm (curve 4 in Fig. 6.6b) $P_\sigma \approx 0.52$ (curve 4 in Fig. 6.7). We emphasize that though the dependences $P_\sigma(\lambda)$ (Fig. 6.7) have resonant character, in no way it can be treated as resonance absorption by an impedance surface of a waveguide iris. It is bound up first of all with range change of emissivity of a slot element, namely the more active a slot excites diffraction fields in jointed electrodynamic volumes the greater amplitudes of these fields, and accordingly, the greater amplitude of currents induced on boundary surfaces. Therefore the curves $P_\sigma(\lambda)$ "iterate" resonant features of the range dependences of the power transmission coefficient P_{10} (Figs. 6.6a, 6.7).

The similar physical regularities are also observed for joint of semi-infinite rectangular waveguides different in dimensions of cross sections. For substantiation of such assertion Figs. 6.8 and 6.9 represent an example of the results of calculations of range characteristics of asymmetrical H-plane waveguide coupling for which $a_{ext} = 46.0$ mm, $b_{ext} = b_{in} = 10.0$ mm and $x_{0ext} = a_{ext}/4$, $y_{0ext} = b_{ext}/4$. The other parameters of the problem and numeration of the curves are chosen in the same

Fig. 6.8 The coefficients of the scattering matrix dependences from the wavelength for an asymmetrical H-plane waveguide coupling at:
$a_{in} = 23.0$ mm,
$a_{ext} = 46.0$ mm,
$b_{in} = b_{ext} = 10.0$ mm,
$d = 1.5$ mm, $2L = 16.0$ mm,
$x_{0in} = a_{in}/2$, $x_{0ext} = a_{ext}/4$,
$y_{0in} = b_{in}/4$, $y_{0ext} = b_{ext}/4$

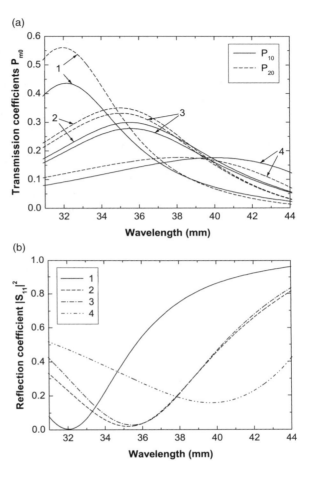

Fig. 6.9 The loss power dependences from the wavelength for an asymmetrical H-plane waveguide coupling at:
$a_{in} = 23.0$ mm,
$a_{ext} = 46.0$ mm,
$b_{in} = b_{ext} = 10.0$ mm,
$d = 1.5$ mm, $2L = 16.0$ mm,
$x_{0in} = a_{in}/2$, $x_{0ext} = a_{ext}/4$,
$y_{0in} = b_{in}/4$, $y_{0ext} = b_{ext}/4$

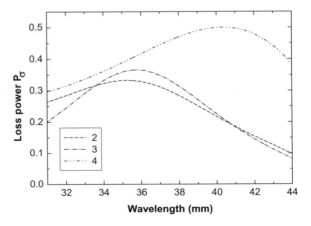

way as for the previous series of calculations. The waveguide device geometry under consideration determines possibility of propagation of two modes, namely, H_{10} and H_{20} (the onset of the H_{30}-wave is excluded by requirements of the excitation) in the waveguide section "*ext*". As one can see from Figs. 6.8 and 6.9, the two-mode operation of the waveguide section "*ext*" really has no effect inherently upon the regularities mentioned above.

A thin layer of a concrete magnetodielectric applied in practice in radiating devices for special purposes [6.4] has been used in the considered examples as realization of an impedance surface at the iris. The analysis of these examples confirms the evident assumption that other materials in which there is practically no microwave loss should be used for control functions in waveguide devices under investigation. In this connection it is expedient to analyze the possibility of influence of a surface impedance of an iris upon redistribution of transmitted power between

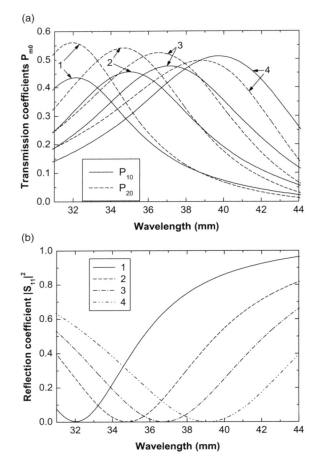

Fig. 6.10 The coefficients of the scattering matrix dependences from the wavelength for an asymmetrical H-plane waveguide coupling at: $a_{in} = 23.0$ mm, $a_{ext} = 46.0$ mm, $b_{in} = b_{ext} = 10.0$ mm, $d = 1.5$ mm, $2L = 16.0$ mm, $x_{0in} = a_{in}/2$, $x_{0ext} = a_{ext}/4$, $y_{0in} = b_{in}/4$, $y_{0ext} = b_{ext}/4$

propagated modes in a waveguide "*ext*" in the case of hypothetically pure imaginary value of an inductive impedance.

Figure 6.10 represents the results of calculations of asymmetrical H-plane of waveguide coupling of the same geometry (see Fig. 6.8) in a similar form for various values of inductive impedances $\overline{Z}_s^{in}$ and $\overline{Z}_s^{ext}$. In the given example curve 1 corresponds to values of impedances $\overline{Z}_s^{in,ext} = 0$, curve 2 corresponds to $\overline{Z}_s^{in} = 0$ and $\overline{Z}_s^{ext} = i0.05$, curve 3 corresponds to $\overline{Z}_s^{in} = i0.02$ and $\overline{Z}_s^{ext} = i0.05$, curve 4 corresponds to equal values of impedances $\overline{Z}_s^{in,ext} = i0.05$. As can be seen from Fig. 6.10, by means of variation of corresponding values of $\overline{Z}_s^{in,ext}$ it is possible to ensure wide-band equal division of the power transmitted in a waveguide section "*ext*" between the H_{10}- and H_{20}- modes in it.

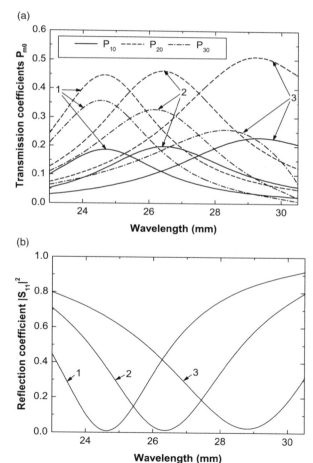

Fig. 6.11 The coefficients of the scattering matrix dependences from the wavelength for an asymmetrical H-plane waveguide coupling at:
$a_{in} = 23.0$ mm,
$a_{ext} = 46.0$ mm,
$b_{in} = b_{ext} = 10.0$ mm,
$d = 1.5$ mm, $2L = 12.0$ mm,
$x_{0in} = a_{in}/2$, $x_{0ext} = a_{ext}/6$,
$y_{0in} = b_{in}/4$, $y_{0ext} = b_{ext}/4$

As the number of propagated modes in the waveguide "*ext*" grows, the analysis of division of transmitted power between them becomes rather complicated. However, one can show that in these cases it is possible also to realize wide-band retuning of a resonance frequency of the waveguide device by means of variation of values of the impedances $\overline{Z}_s^{in,ext}$. Figure 6.11 represents the dependences of the scattering matrix coefficients $P_{m0}(\lambda)$ (Fig. 6.11a) and $|S_{11}|^2(\lambda)$ (Fig. 6.11b) of asymmetrical H-plane waveguide coupling for which $x_{0ext} = a_{ext}/6$ and $2L = 12.0$ mm (the other geometrical parameters are chosen as well as for the previous series of the calculations). Curve 1 in Fig. 6.11 corresponds to the case of perfectly conductive iris surfaces $\overline{Z}_s^{in,ext} = 0$, curve 2 corresponds to the surface impedance values $\overline{Z}_s^{in} = 0$ and $\overline{Z}_s^{ext} = i0.05$, curve 3 corresponds to the equal values of the impedances $\overline{Z}_s^{in,ext} = i0.05$. One can see from Fig. 6.11 that at a change in the corresponding values of surface impedances it is possible to maintain amplitude relations between the modes that are propagated in the waveguide "*ext*" and between which the transmitted power is divided. This maintenance is possible over a wide range of wavelengths.

Thus, based on the numerical investigations of characteristics of asymmetrical H-plane waveguide coupling (the particular cases of realization of a stepped joint), the regularities of diffraction excitation of a slot element made in an impedance iris have been analyzed. The possibility of essential change of an operating (resonant) wavelength of the waveguide device under consideration at variation of the corresponding values of the distributed impedances on iris surfaces has been shown.

References

6.1. Leontovich, M.A.: On the approximate boundary conditions for the electromagnetic field on surfaces of good conductive bodies. Investigations of Radiowave Propagation. Printing House of the Academy of Sciences of the USSR, Moscow-Leningrad (1948) (in Russian)

6.2. Weinstein, L.A.: The Theory of Diffraction and the Factorization Method. Golem Press, Boulder (1969)

6.3. Penkin, Yu.M.: Admittance of slots with coordinate boundaries in a half-infinite rectangular waveguide with impedance endface. Telecommunications and Radio Engineering. **52**, 22–26 (1998)

6.4. Yoshitomi, K.: Radiation from a slot in an impedance surface. IEEE Trans. Antennas and Propagat. **AP-49**, 1370–1376 (2001)

6.5. Takizawa, K., Hashimoto, O.: Transparent wave absorber using resistive thin film at V-band frequency. IEEE Trans. Microwave Theory and Tech. **MTT-47**, 1137–1141 (1999)

6.6. Brekhovskikh, L.M.: Waves in Layered Media. Academic Press, NY (1980)

Chapter 7
Coupling of Some Different Electromagnetic Volumes via Narrow Slots

In the previous chapters we considered the problem of coupling of two electromagnetic volumes via one or more slots. However, more complex structures, representing the cascade coupling of different volumes such as, for example, a waveguide, a resonator, half-space over a perfectly conducting surface, are widely used in practice. Such structures sufficiently widen the possibilities of formation of the required characteristics of antenna-waveguide systems [5.1, 5.2, 7.1–7.4]. In this chapter the method of induced magnetomotive forces is used to solve the problem of magnetic current distribution in an antenna-waveguide structure of the "coupling slots in waveguide wall— rectangular resonator—radiating slot" type for the cases of infinite and semi-infinite rectangular waveguides.

7.1 Problem Formulation and Solution of the Integral Equations

Assume that three electrodynamic volumes with perfectly conducting walls, corresponding to a rectangular waveguide (infinite or semi-infinite) with cross section $\{a \times b\}$ (index "Wg"), a rectangular resonator with dimensions $\{a_R \times b_R \times H\}$ (index "R"), and a half-space above an infinite screen (index "Hs") are linked with one another by rectangular slots S_1 and S_2 cut in infinitely thin common walls (Fig. 7.1). Physical dimensions of the slots (Fig. 7.1c) satisfy the following relations:

$$\frac{d_1}{2L_1} << 1, \quad \frac{d_2}{2L_2} << 1, \quad \frac{d_1}{\lambda} << 1, \quad \frac{d_2}{\lambda} << 1, \qquad (7.1)$$

where $2L_{1,2}$ and $d_{1,2}$ are the lengths and widths of the slots, and λ is the wavelength in vacuum. In this case the equivalent magnetic currents in the slots can be represented in the form ($\vec{e}_{s_{1,2}}$ are the unit vectors of local coordinates tied to the slots):

$$\vec{J}_1(s_1) = \vec{e}_{s_1} J_{01} f_1(s_1) \chi_1(\xi_1), \quad \vec{J}_2(s_2) = \vec{e}_{s_2} J_{02} f_2(s_2) \chi_2(\xi_2), \qquad (7.2)$$

M.V. Nesterenko et al., *Analytical and Hybrid Methods in the Theory of Slot-Hole Coupling of Electrodynamic Volumes*, DOI: 10.1007/978-0-387-76362-0_7,

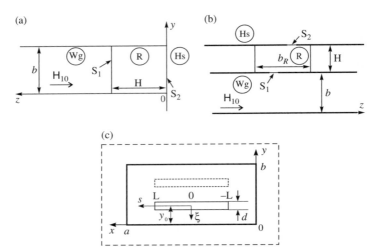

Fig. 7.1 The problem formulation and the symbols used

The functions $f_{1,2}(s_{1,2})$ must satisfy the boundary conditions $f_{1,2}(\pm L_{1,2}) = 0$, while the functions $\chi_{1,2}(\xi_{1,2})$—the normalization conditions $\int_{\xi_{1,2}} \chi_{1,2}(\xi_{1,2})$ $d\xi_{1,2} = 1$.

By using the boundary conditions of continuity of tangential components of the magnetic fields on the surface of every slot, and assuming that the field of impressed sources is concentrated in the "Wg" volume, we come to a system of differential equations in terms of the currents $J_1(s_1)$ and $J_2(s_2)$:

$$
\begin{cases}
\left(\dfrac{d^2}{ds_1^2} + k^2\right)
\begin{bmatrix}
\displaystyle\int_{-L_1}^{L_1} J_1(s_1')\left[G_{s_1}^{Wg}(s_1, s_1') + G_{s_1}^{R}(s_1, s_1')\right] ds_1' \\[2mm]
+ \displaystyle\int_{-L_2}^{L_2} J_2(s_2')G_{s_1}^{R}(s_1, s_2')ds_2'
\end{bmatrix}
= -i\omega H_{0s_1}(s_1), \\[8mm]
\left(\dfrac{d^2}{ds_2^2} + k^2\right)
\begin{bmatrix}
\displaystyle\int_{-L_2}^{L_2} J_2(s_2')\left[G_{s_2}^{R}(s_2, s_2') + G_{s_2}^{Hs}(s_2, s_2')\right] ds_2' \\[2mm]
+ \displaystyle\int_{-L_1}^{L_1} J_1(s_1')G_{s_2}^{R}(s_2, s_1')ds_1'
\end{bmatrix}
= 0.
\end{cases}
\tag{7.3}
$$

Here $G_{s_{1,2}}^{Wg,R,Hs}(s_{1,2}, s_{1,2}')$ are the quasi-one-dimensional Green's functions for the vector potential of the respective volumes, $H_{0s_1}(s_1)$ is the projection of impressed sources field onto the first slot axis.

The functions $f_{1,2}(s_{1,2})$ relating the currents in the slots to the longitudinal coordinates will be selected as follows: the function $\{f_1(s_1)\}$ is obtained from the approximate solution (3.8) of the integral equation for the current in an individual lateral slot excited by the H_{10} type wave and relating to two rectangular waveguides; and the function $\{f_2(s_2)\}$ is derived for a slot in an infinite screen and induced by a planar electromagnetic wave, whose vector $\vec{H}$ is parallel to the vector $\vec{e}_{s_2}$:

$$J_1(s_1) = J_{01}(\cos k s_1 \cos \frac{\pi L_1}{a} - \cos k L_1 \cos \frac{\pi s_1}{a}),$$

$$J_2(s_2) = J_{02}(\cos k s_2 - \cos k L_2).$$

(7.4)

Substituting (7.4) into (7.3) and in conformity with the method of induced MMF for a multi-slot structure, we rearrange equations (7.3) into a system of algebraic equations in terms of unknown amplitudes of the currents J_{01} and J_{02}:

$$\begin{cases} J_{01}\left[Y_1^{Wg}(kd_1, kL_1) + Y_1^{R}(kd_1, kL_1)\right] + J_{02}Y_{12}^{R}(kL_1, kL_2) \\ \qquad\qquad = -\frac{i\omega}{2k}\int\limits_{-L_1}^{L_1} f_1(s_1)H_{0s_1}(s_1)\,ds_1 \\ J_{02}\left[Y_2^{R}(kL_2, kL_2) + Y_2^{Hs}(kd_2, kL_2)\right] + J_{01}Y_{21}^{R}(kL_2, kL_1) = 0. \end{cases}$$

(7.5)

Here the matrix coefficients of the system represent the admittances values of the slots and are given in Appendix C.

Taking into account the fact that for the structure depicted in Fig. 7.1a, we have $H_{0s_1}(s_1) = 2H_0 \cos \frac{\pi s_1}{a}$ while for the structure of Fig.7.1b $H_{0s_1}(s_1) = H_0 \cos \frac{\pi s_1}{a}$, where H_0 is the amplitude of the H_{10} type wave, we can find the required expressions for the amplitudes of currents in each slot:

$$J_{01} = -\left(\frac{i\omega H_0}{\{2\}k^2}\right)\frac{f(kL_1)(Y_2^{R} + Y_2^{Hs})}{(Y_1^{Wg} + Y_1^{R})(Y_2^{R} + Y_2^{Hs}) - Y_{12}^{R}Y_{21}^{R}},$$

$$J_{02} = \left(\frac{i\omega H_0}{\{2\}k^2}\right)\frac{f(kL_1)Y_{21}^{R}}{(Y_1^{Wg} + Y_1^{R})(Y_2^{R} + Y_2^{Hs}) - Y_{12}^{R}Y_{21}^{R}},$$

(7.6)

$$f(kL_1) = 2\cos \frac{\pi L_1}{a}\frac{\sin k L_1 \cos \frac{\pi L_1}{a} - \left(\frac{\pi}{ka}\right)\cos k L_1 \sin \frac{\pi L_1}{a}}{1 - (\pi/ka)^2}$$

$$-\cos k L_1 \frac{\sin \frac{2\pi L_1}{a} + \frac{2\pi L_1}{a}}{(2\pi/ka)},$$

where the multiplier equal to 2 corresponds to the structure of Fig.7.1b. It should be stressed that the choice of the approximating function of the magnetic current distribution in the radiating slot in the form of $f_2(s_2) = (\cos k s_2 - \cos k L_2)$ permitted us to obtain analytical expressions for Y_2^{Hs}, J_{01} and J_{02}.

With the knowledge of the distribution of equivalent magnetic currents in the slots, we can find the energetic characteristics of the cavity-backed slot systems. The reflection coefficient in terms of the field S_{11} and the power radiation coefficient $|S_\Sigma|^2$ for the structure of Fig. 7.1a will equal, respectively,

$$S_{11} = \left\{ 1 - \frac{8\pi k_g f^2 (kL_1)(Y_2^R + Y_2^{Hs})}{i abk^3 [(Y_1^{Wg} + Y_1^R)(Y_2^R + Y_2^{Hs}) - Y_{12}^R Y_{21}^R]} \right\} e^{-i2k_g z}, \tag{7.7}$$

$$|S_\Sigma|^2 = 1 - |S_{11}|, \quad k_g = \sqrt{k^2 - (\pi/a)^2}.$$

For the structure of Fig. 7.1b we have

$$|S_\Sigma|^2 = 1 - |S_{11}^-|^2 - |S_{11}^+|^2, \quad S_{11}^- = (S_{11}^+ - 1)e^{-i2k_g z}, \tag{7.8}$$

where

$$S_{11}^+ = 1 - \frac{2\pi k_g f^2 (kL_1)(Y_2^R + Y_2^{Hs})}{i abk^3 [(Y_1^{Wg} + Y_1^R)(Y_2^R + Y_2^{Hs}) - Y_{12}^R Y_{21}^R]}$$

is the coefficient of transmission over the field.

Let us note that in this chapter the power radiation coefficient $|S_\Sigma|^2$ into the half-space over the perfectly conducting plane will be the main characteristic of the considered slot structures in which we are interested.

7.2 Numerical Results

In Figs. 7.2 and 7.3 we can see the plotted curves of the power radiation coefficient $|S_\Sigma|^2$ as a function of wavelength (frequency responses) in the single-mode range of a standard rectangular waveguide (here and further in this chapter $a = 23.0$ mm, $b = 10.0$ mm) for the system "resonant iris—reentrant resonator – radiating slot" (Fig. 7.1a).

Analysis of the curves permits us to make the following conclusions:

- A solitary slot in an infinitely thin end wall ($H = 0$, $h = 0$) radiates nearly all power supplied, but the quality factor of the resonant curve is rather low.
- If the end wall has a finite thickness ($H = 0$, $h = d$), the resonant frequency changes (towards the short-wave part of the band) together with the frequency response quality. The reason is, that in this case the coupling of the waveguide to the half-space occurs mainly through the cavity resonator formed by the slot cavity.
- Arrangement of the reentrant resonator ($H = a/2$) in the waveguide channel enhances the system quality substantially. In this case the resonant curve has a higher steepness, and its shape tends to rectangular. Obviously, the cascade

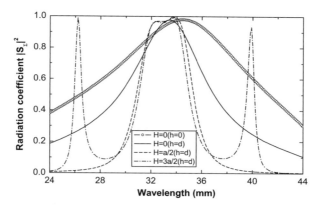

Fig. 7.2 The radiation coefficient dependences from the wavelength for a structure on Fig. 7.1a at: $d_1 = d_2 = d = 1.6\,\text{mm}$, $h_1 = h_2 = h$, $y_{01} = y_{02} = b_R/8$, $2L_1 = 2L_2 = 16.0\,\text{mm}$

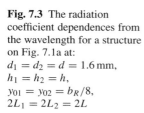

Fig. 7.3 The radiation coefficient dependences from the wavelength for a structure on Fig. 7.1a at: $d_1 = d_2 = d = 1.6\,\text{mm}$, $h_1 = h_2 = h$, $y_{01} = y_{02} = b_R/8$, $2L_1 = 2L_2 = 2L$

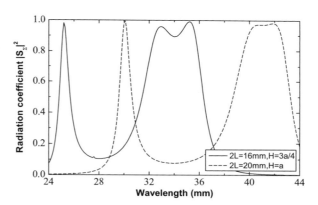

connection of several reentrant resonators will result in a larger change of the frequency response slopes.

- By varying the reentrant resonator length ($H = 3a/4$, $H = 3a/2$) we can attain a maximum of radiation (without regard for the finite conductance of the walls) at two or three frequencies in the band of the H_{10}-wave. The shape of the main resonant curve is almost coincident with the case of $H = a/2$.
- Variation of the slot length and, respectively, of the resonator length results in "shifting" of the extrema of the frequency responses (almost without any change of their shape) to one part of the operating band or another (Fig. 7.3).

For the structure considered above, it is possible, but hardly expedient, to make the resonator cross section other than the waveguide cross section. Figure 7.4 illustrates the effect of changing the resonator's physical dimensions on the relations between the radiation coefficient and the wavelength for the structure "lateral slot in

Fig. 7.4 The radiation coefficient dependences from the wavelength for a structure on Fig. 7.1b at:
$d_1 = d_2 = 1.6$ mm,
$h_1 = h_2 = 1.6$ mm,
$y_{01} = y_{02} = b_R/2$,
$2L_1 = 2L_2 = 16.0$ mm,
$H = a_R/2$

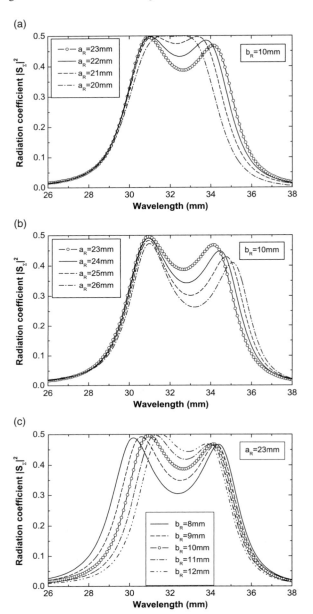

the broad wall of the rectangular waveguide—resonator—radiating slot." As can be seen from the plotted curves, the system with a maximum of quality factor and with the flat test frequency response for slots 16 mm long is produced at the resonator's parameters $a_R = 20.0$ mm, $b_R = 10.0$ mm, and $H = 10.0$ mm (Fig. 7.4a).

By variation of slot locations about the resonator walls, and by increasing their thickness, we can attain a further increase in the quality factor and slopes steepness of the resonant curves (Fig. 7.5).

Fig. 7.5 The radiation coefficient dependences from the wavelength for a structure on Fig. 7.1b at:
$d_1 = d_2 = 1.6\,\text{mm}$,
$h_1 = h_2 = 1.6\,\text{mm}$,
$y_{01} = y_{02} = y_0$,
$2L_1 = 2L_2 = 16.0\,\text{mm}$

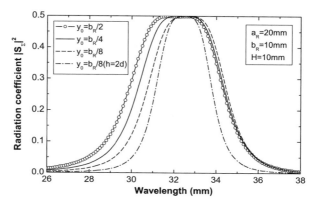

Fig. 7.6 The radiation coefficient dependences from the wavelength for a structure on Fig. 7.1b at: $d_1 = d_2 = h_1 = h_2 = 1.6\,\text{mm}$,
$y_{01} = y_{02} = b_R/2$,
$2L_1 = 2L_2 = 16.0\,\text{mm}$

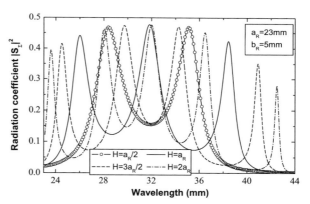

Fig. 7.7 The radiation coefficient dependences from the wavelength for a structure on Fig. 7.1a at:
$d_1 = d_2 = 1.5\,\text{mm}$,
$h_1 = h_2 = 2.0\,\text{mm}$,
$y_{01} = y_{02} = b/4$,
$2L_1 = 2L_2 = 16.0\,\text{mm}$

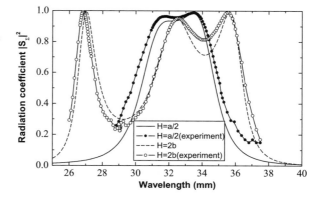

A considerable reduction of resonator height and an increase in its length result in multi-resonance frequency responses in the band of the H_{10}-wave (Fig. 7.6). For example, at the resonator length $H = 2a_R$ the number of resonant peaks is equal to five, but then the level of the radiation factor is higher in the central part of the operating band (i.e., near the resonant wavelengths of the slots about 32.0 mm long) than at its ends.

The comparison of theoretical inquiries with experimental data (for the structure depicted in Fig. 7.1a) confirms the adequacy of the suggested mathematical model to actual physical processes in the cavity-backed slot systems considered (Fig. 7.7).

7.3 Multi-Slot Iris in the Problem of Some Volumes Coupling

One of the methods of possibilities expansion to change antenna-waveguide systems characteristics is to use multi-slot irises in a waveguide [7.5–7.7]. Let us investigate how the electrodynamic characteristics of the structure, given in Fig. 7.1a, will be changed in a case where the iris has two slots.

Let two electrodynamic volumes with perfectly conducting walls, representing a half-infinite rectangular waveguide with the $\{a \times b\}$ cross section (index "Wg"), the rectangular resonator with the $\{a \times b \times H\}$ sizes (index "R") and the half-space above the infinite screen (index "Hs"), respectively, coupled between each other by the rectangular slots S_1, S_2, S_3, cut through in the infinitely thin mutual walls (Fig. 7.8). The geometrical sizes of all slots, as earlier, satisfy the following conditions:

$$\frac{d_m}{2L_m} << 1, \quad \frac{d_m}{\lambda} << 1, m = 1, 2, 3. \tag{7.9}$$

where $2L_m$ and d_m are the lengths and the widths of the slots, correspondingly.

Choosing as the $f_m(s_m)$ functional dependencies from the longitudinal coordinates of the magnetic currents in the slots of function (7.4)

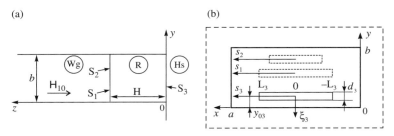

Fig. 7.8 The problem formulation and the symbols used

$$f_{1,2}(s_{1,2}) = (\cos k s_{1,2} \cos \frac{\pi L_{1,2}}{a} - \cos k L_{1,2} \cos \frac{\pi s_{1,2}}{a}), \tag{7.10}$$

$$f_3(s_3) = (\cos k s_3 - \cos k L_3),$$

we obtain the algebraic equations system concerning the unknown amplitude J_{0m} of the currents:

$$
\begin{cases}
J_{01}(Y_{11}^{Wg} + Y_{11}^R) + J_{02}(Y_{12}^{Wg} + Y_{12}^R) + J_{03}Y_{13}^R = -\dfrac{i\omega}{2k} \displaystyle\int_{-L_1}^{L_1} f_1(s_1)H_{0s_1}(s_1)\,ds_1, \\[2mm]
J_{02}(Y_{22}^{Wg} + Y_{22}^R) + J_{01}(Y_{21}^{Wg} + Y_{21}^R) + J_{03}Y_{23}^R = -\dfrac{i\omega}{2k} \displaystyle\int_{-L_2}^{L_2} f_2(s_2)H_{0s_2}(s_2)\,ds_2, \\[2mm]
J_{03}(Y_{33}^R + Y_{33}^{Hs}) + J_{01}Y_{31}^R + J_{02}Y_{32}^R = 0.
\end{cases} \tag{7.11}
$$

The expressions for the Y_{mn}^{Wg}, Y_{mn}^R, Y_{33}^{Hs} admittances are given in Appendix C.

To solve the equations system (7.10), taking into account that for the H_{10} type wave in a half-infinite rectangular waveguide $H_{0s_{1,2}}(s_{1,2}) = 2H_0 \cos \frac{\pi s_{1,2}}{a}$, ($H_0$ is the wave amplitude), we obtain the currents in each of the slots and the reflection coefficient in terms of the field S_{11} and the power radiation coefficient $|S_\Sigma|^2$:

$$S_{11} = \left\{ 1 - \frac{8\pi\gamma}{iabk^3} \left[\tilde{J}_{01} F(kL_1) + \tilde{J}_{02} F(kL_2) \right] \right\} e^{-i2\gamma z}, \tag{7.12}$$

$$|S_\Sigma|^2 = 1 - |S_{11}|^2.$$

Here $\tilde{J}_{0m} = J_{0m} / \left(-\frac{i\omega}{k^2} H_0 \right)$ are the normalized current amplitudes in the slots,

$$F(kL_m) = 2\cos\frac{\pi L_m}{a}\; \frac{\sin k L_m \cos \dfrac{\pi L_m}{a} - \dfrac{\pi}{ka}\cos k L_m \sin \dfrac{\pi L_m}{a}}{1 - (\pi/ka)^2}$$

$$- \cos k L_m \frac{\sin \dfrac{2\pi L_m}{a} + \dfrac{2\pi L_m}{a}}{(2\pi/ka)}.$$

Figures 7.9 to 7.13 give the curves of the radiation coefficient $|S_\Sigma|^2$ dependences from the wavelength in one-mode band of a standard rectangular waveguide for three variants of the waveguide-resonator-slot structure: "1 slot" – the S_3 radiating slot in the end face of the half-infinite rectangular waveguide; "2 slots" – the S_1 slot in the iris and S_3 radiating slot; "3 slots" – the S_1 and S_2 slots in the iris and the S_3 slot. As can be seen, availability of the second slot in the iris leads to the full reflection ($|S_{11}| = 1.0$, $|S_\Sigma|^2 = 0$) of the H_{10} type wave, falling on the iris in the system on a definite length of the λ_{sc} wave, dependent of the $2L_2$ slot length (Fig. 7.9). At this the transmission range at the $0.5|S_\Sigma|^2$ level narrows sufficiently in comparison with the cases of one- and two- slot structures, and the radiation

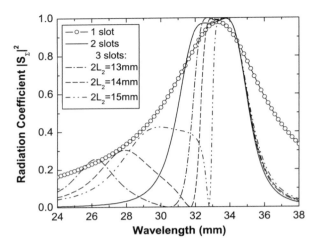

Fig. 7.9 The radiation coefficient dependences from the wavelength for one- and two- slot irises at:
$d_m = h_m = 1.0$ mm,
$2L_1 = 2L_3 = 16.0$ mm,
$H = a/2$, $y_{02} = b/2$,
$y_{01} = y_{03} = b/8$

coefficient increases in the region of short wavelengths, the maximum location (see Table 7.1) of which can be obtained due to formula (5.22). The change of the S_2 slot location in the iris wall leads to conformity to natural laws of the behavior of the system electrodynamic characteristics, analogous to the case of the variation of the length of the $2L_2$ second slot at the fixed value of y_{02} (Fig. 7.10).

The S_3 radiating slot location in the end-face of the waveguide does not practically influence the value and the curve trend $|S_\Sigma|^2 = f(\lambda)$ (Fig. 7.11). Meanwhile both the band of the frequency-energetic characteristics and the λ_{sc} values (Fig. 7.12) change simultaneously at the S_2 slot location near the waveguide broad wall and at relatively-symmetrical rearrangement of the S_1 and S_3 slots, relatively to the $y = b/2$ plane.

One can attempt the maximal radiation on one (Fig. 7.13a) or two wavelengths (Fig. 7.13b) by varying of the H length of the reentrant resonator (Fig. 7.13); at this time the $|S_\Sigma|^2$ second peak appears near the λ_{101} wave resonant length of the resonator (Table 7.1). The value of the width of the system transmission band at the level of $0.5|S_\Sigma|^2$ depends on the length change of the resonator: when H decreases, the band narrows, and when it increases – the band widens, and the λ_{sc} value does not practically depend upon the sizes of the reentrant resonator.

The comparison of the theoretical results with the experimental data for two- and three-slot structures is given in Fig. 7.14.

Table 7.1 The resonant wavelength of a symmetrical coordinate irises and the resonant wavelength of a resonator H_{101}-mode of transmission at: $a = 23.0$ mm, $b = 10.0$ mm, $d_2 = h_2 = 1.0$mm

$y_{02} = b/2$		$2L_2 = 14.0$mm		H, mm	λ_{101}, mm
$2L_2$, mm	λ_{res}, mm	y_{02}	λ_{res}, mm		
13.0	26.00	$b/4$	28.26	11.5	20.57
14.0	28.00	$b/2$	28.00	14.5	24.53
15.0	30.02	$7b/8$	28.87	17.5	27.85
				20.5	30.61

Fig. 7.10 The radiation coefficient dependences from the wavelength for one- and two- slot irises at: $d_m = h_m = 1.0$ mm, $2L_1 = 2L_3 = 16.0$ mm, $2L_2 = 14.0$ mm, $H = a/2$, $y_{01} = y_{03} = b/8$

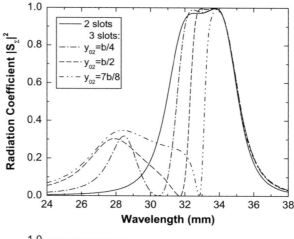

Fig. 7.11 The radiation coefficient dependences from the wavelength for two-slot iris at: $d_m = h_m = 1.0$ mm, $2L_1 = 2L_3 = 16.0$ mm, $2L_2 = 14.0$ mm, $H = a/2$, $y_{01} = b/8$, $y_{02} = b/2$

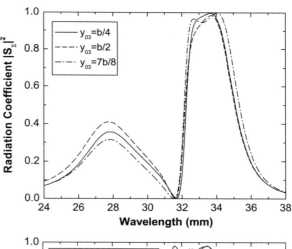

Fig. 7.12 The radiation coefficient dependences from the wavelength for two-slot iris at: $d_m = h_m = 1.0$ mm, $2L_1 = 2L_3 = 16.0$ mm, $2L_2 = 14.0$ mm, $H = a/2$, $y_{02} = b/8$

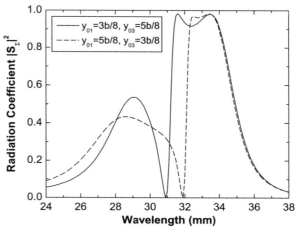

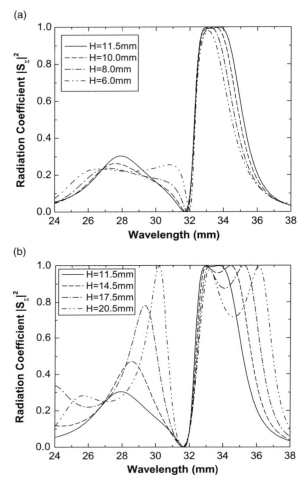

Fig. 7.13 The radiation coefficient dependences from the wavelength for two-slot iris at: $d_m = h_m = 1.0\,\text{mm}$, $2L_1 = 2L_3 = 16.0\,\text{mm}$, $2L_2 = 14.0\,\text{mm}$, $y_{01} = y_{03} = b/8$, $y_{02} = b/2$

Let us note in conclusion that in the case where one (or two) of the iris slots is turned through the $\varphi_{1(2)}$ angle relatively to $\{0x\}$ axis, the distribution function of the current will have the form (5.14):

$$f_{1(2)}(s_{1(2)}) = \cos k s_{1(2)} \cos \frac{\pi L_{1(2)} \cos \varphi_{1(2)}}{a} - \cos k L_{1(2)} \cos \frac{\pi s_{1(2)} \cos \varphi_{1(2)}}{a}.$$

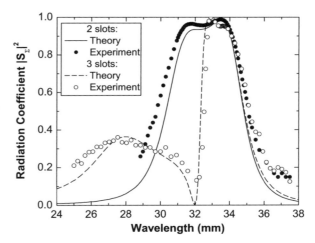

Fig. 7.14 The radiation coefficient dependences from the wavelength for one- and two- slot irises at: $d_m = 1.5$ mm, $h_m = 2.0$ mm, $2L_1 = 2L_3 = 16.0$ mm, $2L_2 = 14.0$ mm, $H = a/2$, $y_{01} = y_{03} = b/4$, $y_{02} = 3b/4$

References

7.1. Chen, T.S.: Characteristics of waveguide resonant-iris filters. IEEE Trans. Microwave Theory and Tech. **MTT-15**, 260–262 (1967)

7.2. Long, S.A.: Experimental study of the impedance of cavity-backed slot antennas. IEEE Trans. Antennas and Propag. **AP-23**, 1–7 (1975)

7.3. Lee, J.Y., Horng, T.Sh., Alexopoulos, N.G.: Analysis of cavity-backed aperture antennas with a dielectric overlay. IEEE Trans. Antennas and Propag. **AP-42**, 1556–1562 (1994)

7.4. Nesterenko, M.V., Katrich, V.A.: The method of induced magnetomotive forces for cavity-backed slot radiators and coupling slots. Radioelectronics and Communications Systems. **47**, 8–13 (2004)

7.5. Paterson, N.G., Anderson, I.: Bandstop iris for rectangular waveguide. Electronics Letters. **12**, 592–594 (1976)

7.6. Kirilenko, A.A., Mospan, L.P.: Reflection resonances and natural frequencies of two-aperture iris in rectangular waveguide. IEEE Trans. Microwave Theory and Tech. **MTT-48**, 1419–1421 (2000)

7.7. Kirilenko, A.A., Mospan, L.P.: Two- and three- slot irises as the bandstop filter sections. Microwave and Opt. Tech. Letters. **28**, 282–284 (2001)

Appendix A
Magnetic Dyadic Green's Functions of the Considered Electrodynamic Volumes

1. Infinite rectangular waveguide with perfectly conducting walls:

$$
\hat{G}^m(\vec{r}, \vec{r}') = \frac{2\pi}{ab} \sum_{m,n} \frac{\varepsilon_m \varepsilon_n}{k_z} \Big\{ (\vec{e}_x \otimes \vec{e}_{x'}) \Phi_x^m(x, y; x', y') e^{-k_z|z-z'|}
$$
$$
+ (\vec{e}_y \otimes \vec{e}_{y'}) \Phi_y^m(x, y; x', y') e^{-k_z|z-z'|} \tag{A.1}
$$
$$
+ (\vec{e}_z \otimes \vec{e}_{z'}) \Phi_z^m(x, y; x', y') e^{-k_z|z-z'|} \Big\} .
$$

2. Semi-infinite rectangular waveguide with perfectly conducting walls:

$$
\hat{G}^m(\vec{r}, \vec{r}') = \frac{2\pi}{ab} \sum_{m,n} \frac{\varepsilon_m \varepsilon_n}{k_z} \Big\{ (\vec{e}_x \otimes \vec{e}_{x'}) \Phi_x^m(x, y; x', y') \Big[e^{-k_z|z-z'|} + e^{-k_z(z+z')} \Big]
$$
$$
+ (\vec{e}_y \otimes \vec{e}_{y'}) \Phi_y^m(x, y; x', y') \Big[e^{-k_z|z-z'|} + e^{-k_z(z+z')} \Big] \tag{A.2}
$$
$$
+ (\vec{e}_z \otimes \vec{e}_{z'}) \Phi_z^m(x, y; x', y') \Big[e^{-k_z|z-z'|} - e^{-k_z(z+z')} \Big] \Big\} .
$$

3. Rectangular resonator with perfectly conducting walls:

$$
\hat{G}^m(\vec{r}, \vec{r}') = \frac{2\pi}{a_R b_R} \sum_{m,n} \frac{\varepsilon_m \varepsilon_n}{k_z}
$$
$$
\times \Big\{ (\vec{e}_x \otimes \vec{e}_{x'}) \Phi_x^m(x, y; x', y') \left[\frac{\mathrm{ch}k_z(h - |z - z'|) + \mathrm{ch}k_z(h - |z + z'|)}{\mathrm{sh}k_z h} \right]
$$
$$
+ (\vec{e}_y \otimes \vec{e}_{y'}) \Phi_y^m(x, y; x', y') \left[\frac{\mathrm{ch}k_z(h - |z - z'|) + \mathrm{ch}k_z(h - |z + z'|)}{\mathrm{sh}k_z h} \right]
$$
$$
+ (\vec{e}_z \otimes \vec{e}_{z'}) \Phi_z^m(x, y; x', y') \left[\frac{\mathrm{ch}k_z(h - |z - z'|) - \mathrm{ch}k_z(h - |z + z'|)}{\mathrm{sh}k_z h} \right] \Big\} .
$$
$$
\tag{A.3}
$$

4. Semi-infinite rectangular waveguide with impedance end in the case where impressed sources are located on the end-wall surface:

$$\hat{G}^M(\vec{r}, \vec{r}') = \frac{2\pi}{ab} \sum_{m,n} \frac{\varepsilon_m \varepsilon_n}{k_z} \left\{ (\vec{e}_x \otimes \vec{e}_{x'}) \Phi_x^m(x, y; x', y') f_{II}(k_z, \overline{Z}_s) 2e^{-k_z z} + \right.$$
$$\left. + (\vec{e}_y \otimes \vec{e}_{y'}) \Phi_y^m(x, y; x', y') f_{II}(k_z, \overline{Z}_s) 2e^{-k_z z} \right\},$$

(A.4)

where

$$f_{II}(k_z, \overline{Z}_s) = \frac{kk_z(1 + \overline{Z}_s^2)}{(ik + k_z \overline{Z}_s)(k \overline{Z}_s - ik)}.$$

5. Semi-infinite rectangular waveguide with impedance end excited by longitudinal impressed currents:

$$\hat{G}^M(\vec{r}, \vec{r}') = \frac{2\pi}{ab} \sum_{m,n} \frac{\varepsilon_m \varepsilon_n}{k_z} \left\{ (\vec{e}_z \otimes \vec{e}_{z'}) \Phi_z^m(x, y; x', y') \left[\begin{array}{c} e^{-k_z|z-z'|} \\ -f_\perp(k_z, \overline{Z}_s)e^{-k_z(z+z')} \end{array} \right] \right\},$$

(A.5)

where

$$f_\perp(k_z, \overline{Z}_s) = \frac{ik - k_z \overline{Z}_s}{ik + k_z \overline{Z}_s}.$$

6. Half-space over the infinite perfect conducting plane:

$$\hat{G}^m(\vec{r}, \vec{r}') = \hat{I}\frac{e^{-ikR}}{R} + (\vec{e}_x \otimes \vec{e}_{x'})\frac{e^{-ikR_0}}{R_0} + (\vec{e}_y \otimes \vec{e}_{y'})\frac{e^{-ikR_0}}{R_0}$$
$$- (\vec{e}_z \otimes \vec{e}_{z'})\frac{e^{-ikR_0}}{R_0},$$
$$R = \sqrt{(x - x')^2 + (y - y')^2 + (z - z')^2},$$
$$R_0 = \sqrt{(x - x')^2 + (y - y')^2 + (z + z')^2}.$$

(A.6)

The following notations are adopted in expressions (A.1)–(A.6):

$$\Phi_x^m(x, y; x', y') = \sin k_x x \sin k_x x' \cos k_y y \cos k_y y',$$
$$\Phi_y^m(x, y; x', y') = \cos k_x x \cos k_x x' \sin k_y y \sin k_y y',$$
$$\Phi_z^m(x, y; x', y') = \cos k_x x \cos k_x x' \cos k_y y \cos k_y y',$$

$$\varepsilon_{m,n} = \left\{ \begin{array}{ll} 1, & m,n = 0 \\ 2, & m,n \neq 0 \end{array} \right., \quad k_x = \frac{m\pi}{a(a_R)}, \quad k_y = \frac{n\pi}{b(b_R)}, \quad k_z = \sqrt{k_x^2 + k_y^2 - k^2},$$

m and n are integers, $\overline{Z}_s$ is the normalized surface impedance, $\vec{e}_x$, $\vec{e}_y$, and $\vec{e}_z$ are the unit vectors of the Cartesian coordinate system fixed to the waveguide, and "$\otimes$" stands for dyadic product.

Appendix B
Functions of the Own Field of Single Slots

The functions of the slot own field for the case of coupling of two infinite rectangular waveguides are given below:

$$W_t^s(kd, kL) = \frac{8\pi}{ab} \sum_{m,n} \frac{\varepsilon_n e^{-k_z d/4}}{k_z(k^2 - k_x^2)} \sin^2 \frac{m\pi}{2}$$
$$\times \cos k_x L \left(k \sin kL \cos k_x L - k_x \cos kL \sin k_x L \right),$$

(B.1)

$$W_t^a(kd, kL) = -\frac{8\pi}{ab} \sum_{m,n} \frac{\varepsilon_n e^{-k_z d/4}}{k_z(k^2 - k_x^2)} \cos^2 \frac{m\pi}{2}$$
$$\times \sin k_x L \left(k \cos kL \sin k_x L - k_x \sin kL \cos k_x L \right),$$

(B.2)

$$W_{lb}^s(kd, kL) = \frac{4\pi}{ab} \sum_{m,n} \frac{\varepsilon_m \varepsilon_n \cos k_x x_0 \cos k_x \left(x_0 + d/4 \right)}{k_z(k_x^2 + k_y^2)}$$
$$\times e^{-k_z L} \left[k_z \cos kL \, \text{sh} k_z L + k \sin kL \, \text{ch} k_z L \right],$$

(B.3)

$$W_{lb}^a(kd, kL) = \frac{4\pi}{ab} \sum_{m,n} \frac{\varepsilon_m \varepsilon_n \cos k_x x_0 \cos k_x \left(x_0 + d/4 \right)}{k_z(k_x^2 + k_y^2)}$$
$$\times e^{-k_z L} \left[k_z \sin kL \, \text{ch} k_z L - k \cos kL \, \text{sh} k_z L \right],$$

(B.4)

$$W_{ln}^s(kd, kL) = \frac{4\pi}{ab} \sum_{n,n} \frac{\varepsilon_m \varepsilon_n \cos k_y y_0 \cos k_y \left(y_0 + d/4 \right)}{k_z(k_x^2 + k_y^2)}$$
$$\times e^{-k_z L} \left[k_z \cos kL \, \text{sh} k_z L + k \sin kL \, \text{ch} k_z L \right],$$

(B.5)

$$W_{ln}^a\,(kd, kL) = \frac{4\pi}{ab} \sum_{m,n} \frac{\varepsilon_m \varepsilon_n \cos k_y y_0 \cos k_y \left(y_0 + \frac{d}{4}\right)}{k_z(k_x^2 + k_y^2)}$$

$$\times\, e^{-k_z L}\left[k_z \sin kL \mathrm{ch} k_z L - k \cos kL \mathrm{sh} k_z L\right]. \tag{B.6}$$

In the (B.1)–(B.6) expressions there are the following symbols: $k_x = \frac{m\pi}{a}$, $k_y = \frac{n\pi}{b}$, $k_z = \sqrt{k_x^2 + k_y^2 - k^2}$, (m, n = 0, 1, 2 …); $\varepsilon_{m,n} = 1$ at m, n = 0, $\varepsilon_{m,n} = 2$ at m, n $\neq$ 0; x_0 and y_0 are the slots axes coordinates.

The functions of the slot own field for the case of resonant iris with the slot arbitrary oriented in a rectangular waveguide are given below:

$$W_0(kd_e, 2kL) = \frac{4\pi}{ab} \sum_{m=0}^{\infty} \sum_{n=0}^{\infty} \frac{\varepsilon_m \varepsilon_n}{k_z}$$

$$\times \Big\langle \cos^2 \varphi \{\sin kL[\Phi_s(k_x x_0)\Phi_c(k_y y_0)(\cos k_1 L + \cos k_2 L)I_1^+(k_{1,2}L)$$

$$+ \Phi_c(k_x x_0)\Phi_s(k_y y_0)(\cos k_1 L - \cos k_2 L)I_1^-(k_{1,2}L)]$$

$$+ \frac{1}{2}\cos kL \sin 2k_x x_0[\Phi_s(k_y y_0)(\cos k_1 L - \cos k_2 L)I_2^-(k_{1,2}L)$$

$$- \Phi_c(k_y y_0)(\cos k_1 L + \cos k_2 L)I_2^+(k_{1,2}L)]$$

$$+ \frac{1}{2}\cos kL \sin 2k_y y_0[\Phi_s(k_x x_0)(\cos k_1 L + \cos k_2 L)I_2^-(k_{1,2}L)$$

$$- \Phi_c(k_x x_0)(\cos k_1 L - \cos k_2 L)I_2^+(k_{1,2}L)]\}$$

$$+ \sin^2 \varphi \{\sin kL[\Phi_s(k_x x_0)\Phi_c(k_y y_0)(\cos k_1 L - \cos k_2 L)I_1^-(k_{1,2}L)$$

$$+ \Phi_c(k_x x_0)\Phi_s(k_y y_0)(\cos k_1 L + \cos k_2 L)I_1^+(k_{1,2}L)]$$

$$+ \frac{1}{2}\cos kL \sin 2k_x x_0[\Phi_s(k_y y_0)(\cos k_1 L + \cos k_2 L)I_2^+(k_{1,2}L)$$

$$- \Phi_c(k_y y_0)(\cos k_1 L - \cos k_2 L)I_2^-(k_{1,2}L)]$$

$$+ \frac{1}{2}\cos kL \sin 2k_y y_0[\Phi_s(k_x x_0)(\cos k_1 L - \cos k_2 L)I_2^+(k_{1,2}L)$$

$$- \Phi_c(k_x x_0)(\cos k_1 L + \cos k_2 L)I_2^-(k_{1,2}L)]\}$$

$$+ \tfrac{1}{2}\sin kL \sin 2k_x x_0 \sin 2k_y y_0[\cos k_1 L\, I_1(k_1 L) - \cos k_2 L\, I_1(k_2 L)]\Big\rangle, \tag{B.7}$$

$$W_\varphi(kd_e, kL) = \frac{2\pi}{ab} \sum_{m=0}^{\infty} \sum_{n=0}^{\infty} \frac{\varepsilon_m \varepsilon_n}{k_z}$$

$$\times \{[\Phi_s(k_x x_0)\Phi_c(k_y y_0)[(\cos k_1 L + \cos 2\varphi \cos k_2 L)I_1(k_1 L)$$

$$+ (\cos k_2 L + \cos 2\varphi \cos k_1 L)I_1(k_2 L)] \tag{B.8}$$

$$- \Phi_c(k_x x_0)\Phi_s(k_y y_0)[(\cos k_1 L - \cos 2\varphi \cos k_2 L)I_1(k_1 L)$$

$$+ (\cos k_2 L - \cos 2\varphi \cos k_1 L)I_1(k_2 L)]\},$$

$$W(kd_e, kL) = \frac{16\pi}{ab} \sum_{m=1,3\ldots}^{\infty} \sum_{n=0}^{\infty} \frac{\varepsilon_n}{k_z} \cos k_y y_0 \cos k_y \left(y_0 + \frac{d_e}{4} \right) \cos k_x L$$

$$\times \left[\frac{k \sin kL \cos k_x L - k_x \cos kL \sin k_x L}{k^2 - k_x^2} \right].$$

(B.9)

In the (B.7)–(B.9) expressions there are the following symbols:

$$I_1(k_{1,2}L) = \frac{k \sin kL \cos k_{1,2}L - k_{1,2} \cos kL \sin k_{1,2}L}{k^2 - k_{1,2}^2},$$

$$I_2(k_{1,2}L) = \frac{k_{1,2} \sin kL \cos k_{1,2}L - k \cos kL \sin k_{1,2}L}{k^2 - k_{1,2}^2},$$

$$\Phi_s(k_x x_0) = \sin(k_x x_0^{\varphi}) \sin(k_x x_0), \quad \Phi_c(k_x x_0) = \cos(k_x x_0^{\varphi}) \cos(k_x x_0),$$

$$\Phi_s(k_y y_0) = \sin(k_y y_0^{\varphi}) \sin(k_y y_0), \quad \Phi_c(k_y y_0) = \cos(k_y y_0^{\varphi}) \cos(k_y y_0),$$

$$k_{1,2} = k_x \cos \varphi \pm k_y \sin \varphi, \quad x_0^{\varphi} = x_0 - \frac{d_e}{4} \sin \varphi, \quad y_0^{\varphi} = y_0 + \frac{d_e}{4} \cos \varphi,$$

$$I_{1,2}^{\pm}(k_{1,2}L) = I_{1,2}(k_1 L) \pm I_{1,2}(k_2 L).$$

The rest symbols are the same as in the (B.1)–(B.6) formulas.

Appendix C
Eigen and Mutual Slot Admittances

The eigen longitudinal slots admittances:

$$
Y^s(kd, kL) = \frac{2\pi}{ab} \sum_{m=0}^{\infty} \sum_{n=0}^{\infty} \frac{\varepsilon_m \varepsilon_n}{k} \cos k_x x_0 \cos k_x \left(x_0 + \frac{d}{4} \right)
$$
$$
\times \left\{ \left[\cos \gamma L \left(\frac{k}{k_z} \sin kL - \cos kL \right) \right] F^s(k_z L) \right.
$$
$$
- \frac{\cos kL}{k_z^2 + \gamma^2} \left[(k_z^2 + k^2) \left(\frac{\gamma}{k_z} \sin \gamma L - \cos \gamma L \right) F^s(k_z L) \right.
$$
$$
\left. \left. + \left(\frac{\pi}{a} \right)^2 F^s(kL) \right] \right\},
$$

(C.1)

$$
F^s(kL) = 2 \cos \gamma L \, \frac{k \sin kL \cos \gamma L - \gamma \cos kL \sin \gamma L}{(\pi/a)^2}
$$
$$
- \cos kL \, \frac{\sin 2\gamma L + 2\gamma L}{2\gamma},
$$

(C.2)

$$
F^s(k_z L) = \frac{\cos \gamma L}{k_z^2 + k^2} \left[k_z \cos kL \left(1 - e^{-2k_z L} \right) + k \sin kL \left(1 + e^{-2k_z L} \right) \right]
$$
$$
- \frac{\cos kL}{k_z^2 + \gamma^2} \left[k_z \cos \gamma L \left(1 - e^{-2k_z L} \right) \right.
$$
$$
\left. + \gamma \sin \gamma L \left(1 + e^{-2k_z L} \right) \right].
$$

(C.3)

$$
\begin{aligned}
Y^a(kd, kL) = {} & \frac{2\pi}{ab} \sum_{m=0}^{\infty} \sum_{n=0}^{\infty} \frac{\varepsilon_m \varepsilon_n}{k} \cos k_x x_0 \cos k_x \left(x_0 + \frac{d}{4} \right) \\
& \times \left\{ \left[-\sin \gamma L \left(\frac{k}{k_z} \cos kL + \sin kL \right) \right] F^a(k_z L) \right. \\
& + \frac{\sin kL}{k_z^2 + \gamma^2} \left[(k_z^2 + k^2) \left(\frac{\gamma}{k_z} \cos \gamma L + \sin \gamma L \right) F^a(k_z L) \right. \\
& \left. \left. + \left(\frac{\pi}{a} \right)^2 F^a(kL) \right] \right\},
\end{aligned}
\tag{C.4}
$$

$$
\begin{aligned}
F^a(kL) = {} & 2 \sin \gamma L \frac{k \cos kL \sin \gamma L - \gamma \sin kL \cos \gamma L}{(\pi/a)^2} \\
& - \sin kL \frac{\sin 2\gamma L - 2\gamma L}{2\gamma},
\end{aligned}
\tag{C.5}
$$

$$
\begin{aligned}
F^a(k_z L) = {} & \frac{\sin k_g L}{k_z^2 + k^2} \left[k_z \sin kL \left(1 + e^{-2k_z L} \right) - k \cos kL \left(1 - e^{-2k_z L} \right) \right] - \\
& - \frac{\sin kL}{k_z^2 + \gamma^2} \left[k_z \sin \gamma L \left(1 + e^{-2k_z L} \right) - \gamma \cos \gamma L \left(1 - e^{-2k_z L} \right) \right].
\end{aligned}
\tag{C.6}
$$

The mutual longitudinal slots admittances:

$$
\begin{aligned}
Y^{ss}_{\left\{ \substack{12 \\ 21} \right\}}(kL_m, kL_n) = {} & \frac{2\pi}{ab} \sum_{m=0}^{\infty} \sum_{n=0}^{\infty} \frac{\varepsilon_m \varepsilon_n}{kk_z} \mathrm{ch} k_z z_0 \cos k_x x_{0m} \cos k_x (x_{0n} + \frac{d_n}{4}) \\
& \times \left\{ \cos \gamma L_n (k \sin kL_n - k_z \cos kL_n) F^s(k_z L_m) \right. \\
& - \frac{\cos kL_n}{k_z^2 + \gamma^2} [k_z k_c^2 F^s(kL_m)] \left\{ \begin{array}{c} 1 \\ \cos \gamma z_0 \\ \mathrm{ch} k_z z_0 \end{array} \right\} \\
& \left. + (k_z^2 + k^2)(\gamma \sin \gamma L_n - k_z \cos \gamma L_n) F^s(k_z L_m) \right\},
\end{aligned}
\tag{C.7}
$$

$$
\begin{aligned}
Y^{as}_{\left\{ \substack{12 \\ 21} \right\}}(kL_m, kL_n) = {} & \frac{2\pi}{ab} \sum_{m=0}^{\infty} \sum_{n=0}^{\infty} \frac{\varepsilon_m \varepsilon_n}{kk_z} \mathrm{sh} k_z z_0 \cos k_x x_{0m} \cos k_x (x_{0n} + \frac{d_n}{4}) \\
& \times \left\{ \mp \cos \gamma L_n (k \sin kL_n - k_z \cos kL_n) F^a(k_z L_m) \right. \\
& + \frac{\cos kL_n}{k_z^2 + \gamma^2} [k_z k_c^2 F^a(kL_m)] \left\{ \begin{array}{c} \gamma/k_z \\ -\sin \gamma z_0 \\ \mathrm{sh} k_z z_0 \end{array} \right\} \\
& \left. \pm (k_z^2 + k^2)(\gamma \sin \gamma L_n - k_z \cos \gamma L_n) F^a(k_z L_m) \right\},
\end{aligned}
\tag{C.8}
$$

$$Y^{aa}_{\left\{\begin{smallmatrix}12\\21\end{smallmatrix}\right\}}(kL_m, kL_n) = \frac{2\pi}{ab} \sum_{m=0}^{\infty} \sum_{n=0}^{\infty} \frac{\varepsilon_m \varepsilon_n}{kk_z} \mathrm{ch} k_z z_0 \cos k_x x_{0m} \cos k_x (x_{0n} + \frac{d_n}{4})$$
$$\times \{ -\sin \gamma L_n (k \cos kL_n + k_z \sin kL_n) F^a (k_z L_m)$$
$$+ \frac{\sin kL_n}{k_z^2 + \gamma^2} [k_z k_c^2 F^a (kL_m)] \left\{ \begin{matrix} 1 \\ \cos \gamma z_0 \\ \mathrm{ch} k_z z_0 \end{matrix} \right\}$$
$$+ (k_z^2 + k^2)(\gamma \cos \gamma L_n + k_z \sin \gamma L_n) F^a (k_z L_m)]\}, \quad (C.9)$$

$$Y^{sa}_{\left\{\begin{smallmatrix}12\\21\end{smallmatrix}\right\}}(kL_m, kL_n) = \frac{2\pi}{ab} \sum_{m=0}^{\infty} \sum_{n=0}^{\infty} \frac{\varepsilon_m \varepsilon_n}{kk_z} \mathrm{sh} k_z z_0 \cos k_x x_{0m} \cos k_x (x_{0n} + \frac{d_n}{4})$$
$$\times \{ \pm \sin \gamma L_n (k \cos kL_n + k_z \sin kL_n) F^s (k_z L_m)$$
$$- \frac{\sin kL_n}{k_z^2 + \gamma^2} [k_z k_c^2 F^s (kL_m)] \left\{ \begin{matrix} -\gamma / k_z \\ \sin \gamma z_0 \\ \mathrm{sh} k_z z_0 \end{matrix} \right\}$$
$$\pm (k_z^2 + k^2)(\gamma \cos \gamma L_n + k_z \sin \gamma L_n) F^s (k_z L_m)]\}, \quad (C.10)$$

$$Y^s_{mn}(kL) = \frac{2\pi}{ab} \sum_{m=0}^{\infty} \sum_{n=0}^{\infty} \frac{\varepsilon_m \varepsilon_n}{kk_z} \cos k_x x_{0m} \cos k_x (x_{0n} + \frac{d_n}{4})$$
$$\times \{ \cos \gamma L (k \sin kL - k_z \cos kL) F^s (k_z L)$$
$$- \frac{\cos kL}{k_z^2 + \gamma^2} [(k_z / k) k_c^2 f^s (kL)$$
$$+ (k_z^2 + k^2)(\gamma \sin \gamma L - k_z \cos \gamma L) F^s (k_z L)]\}, \quad (C.11)$$

$$Y^a_{mn}(kL) = \frac{2\pi}{ab} \sum_{m=0}^{\infty} \sum_{n=0}^{\infty} \frac{\varepsilon_m \varepsilon_n}{kk_z} \cos k_x x_{0m} \cos k_x (x_{0n} + \frac{d_n}{4})$$
$$\times \{ -\sin \gamma L (k \cos kL + k_z \sin kL) F^a (k_z L)$$
$$+ \frac{\sin kL}{k_z^2 + \gamma^2} [(k_z / k) k_c^2 f^a (kL)$$
$$- (k_z^2 + k^2)(\gamma \cos \gamma L + k_z \sin \gamma L) F^a (k_z L)]\}, \quad (C.12)$$

$$f^s (kL_m) = 2 \cos \gamma L_m \frac{\sin kL_m \cos \gamma L_m - (\gamma / k) \cos kL_m \sin \gamma L_m}{1 - (\gamma / k)^2}$$
$$- \cos kL_m \frac{\sin 2\gamma L_m + 2\gamma L_m}{2(\gamma / k)}, \quad (C.13)$$

$$f^a(kL_m) = 2 \sin \gamma L_m \frac{\cos kL_m \sin \gamma L_m - (\gamma/k) \sin kL_m \cos \gamma L_m}{1 - (\gamma/k)^2}$$
$$- \sin kL_m \frac{\sin 2\gamma L_m - 2\gamma L_m}{2(\gamma/k)}. \tag{C.14}$$

The eigen ($m = n$) and mutual ($m \neq n$) slot admittances of the transverse slots system:

$$Y_{mm}^{Wg}(kL_m, kL_m) = \frac{2\pi}{ab} \sum_{m=1,3...}^{\infty} \sum_{n=0}^{\infty} \frac{\varepsilon_n(k^2 - k_x^2)}{kk_z} e^{-k_z \frac{d_m}{4}} I_{Wg}^2(kL_m), \tag{C.15}$$

$$Y_{mn}^{Wg}(kL_m, kL_n)$$
$$= \frac{2\pi}{ab} \sum_{m=1,3...}^{\infty} \sum_{n=0}^{\infty} \frac{\varepsilon_n(k^2 - k_x^2)}{kk_z} e^{-k_z|z_m - z_n|} I_{Wg}(kL_m) I_{Wg}(kL_n), \tag{C.16}$$

$$Y_{\{^{mm}_{mn}\}}^{R}(kL_m) = \frac{2\pi}{L_m d_m} \sum_{m=1,3...}^{\infty} \sum_{n=0}^{\infty} \frac{\varepsilon_n(k^2 - k_{xR}^2)}{kk_{zR}} \left\{ \begin{array}{c} \coth(k_{zR}h) \\ 1/\mathrm{sh}(k_{zR}h) \end{array} \right\}$$
$$\times \cos \frac{n\pi}{2} \cos \left(k_{yR} \frac{3d_m}{4} \right) I_R^2(kL_m), \tag{C.17}$$

$$I_{Wg(R)}(kL_m)$$
$$= 2 \left\{ \frac{k \sin kL_m \cos k_{x(xR)}L_m - k_{x(xR)} \cos kL_m \sin k_{x(xR)}L_m}{k^2 - k_{x(xR)}^2} \cos k_c L_m \right.$$
$$\left. - \frac{k_c \sin k_c L_m \cos k_{x(xR)}L_m - k_{x(xR)} \cos k_c L_m \sin k_{x(xR)}L_m}{k_c^2 - k_{x(xR)}^2} \cos kL_m \right\}. \tag{C.18}$$

In the (C.15)–(C.18) expressions are the following symbols: $k_{xR} = \frac{m\pi}{2L_m}$, $k_{yR} = \frac{n\pi}{d_m}$, $k_{zR} = \sqrt{k_{xR}^2 + k_{yR}^2 - k^2}$. The rest symbols are the same as in the (B.1)–(B.6) formulas.

The slot admittances for the case of impedance slotted iris:

$$Y^{in,ext}(KL, \overline{Z}_s^{in,ext}) = \frac{4\pi}{a_{in,ext} b_{in,ext}}$$

$$\times \sum_{m=1}^{\infty} \sum_{n=0}^{\infty} \frac{\varepsilon_n (k^2 - k_{x_{in,ext}}^2)}{kk_{z_{in,ext}}} \sin^2(k_{x_{in,ext}} x_{0in,ext}) \qquad \text{(C.19)}$$

$$\times \cos(k_{y_{in,ext}} y_{0in,ext}) \cos\left(k_{y_{in,ext}}\left(y_{0in,ext} + \frac{d_e}{4}\right)\right) F(k_{z_{in,ext}} \overline{Z}_s^{in,ext})$$

$$\times \left[I_{in,ext}(kL) \cos\frac{\pi L}{a_{in}} - I_{in,ext}\left(\frac{\pi L}{a_{in}}\right) \cos kL \right]^2,$$

$$F(k_{z_{in,ext}}, \overline{Z}_s^{in,ext}) = \frac{kk_{z_{in,ext}}(1 + (\overline{Z}_s^{in,ext})^2)}{(ik + k_{z_{in,ext}}\overline{Z}_s^{in,ext})(k\overline{Z}_s^{in,ext} - ik_{z_{in,ext}})}$$

$$\times \left(1 - i\frac{kk_{z_{in,ext}}\overline{Z}_s^{in,ext}}{k^2 - k_{x_{in,ext}}^2}\right), \qquad \text{(C.20)}$$

$$I_{in,ext}(kL) = 2\frac{k \sin kL \cos k_{x_{in,ext}} L - k_{x_{in,ext}} \cos kL \sin k_{x_{in,ext}} L}{k^2 - k_{x_{in,ext}}^2}, \qquad \text{(C.21)}$$

$$I_{in,ext}\left(\frac{\pi L}{a_{in}}\right) = 2\frac{\frac{\pi}{a_{in}} \sin\frac{\pi L}{a_{in}} \cos k_{x_{in,ext}} L - k_{x_{in,ext}} \cos\frac{\pi L}{a_{in}} \sin k_{x_{in,ext}} L}{(\pi/a_{in})^2 - k_{x_{in,ext}}^2}. \qquad \text{(C.22)}$$

In the (C.19)–(C.22) expressions there are the following symbols: $k_{x_{in,ext}} = \frac{m\pi}{a_{in,ext}}$, $k_{y_{in,ext}} = \frac{n\pi}{b_{in,ext}}$, $k_{z_{in,ext}} = \sqrt{k_x^2 + k_y^2 - k^2}$, (m, n = 0, 1, 2...); $\varepsilon_n = 1$ at n = 0, $\varepsilon_n = 2$ at n $\neq$ 0; $x_{0in,ext}$ and $y_{0in,ext}$ are the slots axes coordinates.

The eigen ($m = n$) and mutual ($m \neq n$) slot admittances for the case of coupling of some different volumes:

$$Y_{mn}^{Wg}(kL_m, kL_n) = \frac{4\pi}{ab} \sum_{m=1,3...}^{\infty} \sum_{n=0}^{\infty} \frac{\varepsilon_n (k^2 - k_x^2)}{kk_z}$$

$$\times \cos k_y y_{0m} \cos k_y \left(y_{0n} + \frac{d_n}{4}\right) I_1(kL_m) I_1(kL_n) \qquad \text{(C.23)}$$

–for the structure on Fig. 7.1a,

$$Y_{mm}^{Wg}(kL_m, kL_m) = \frac{2\pi}{ab} \sum_{m=1,3...}^{\infty} \sum_{n=0}^{\infty} \frac{\varepsilon_n(k^2 - k_x^2)}{kk_z} e^{-k_z\frac{d_m}{4}} I_1^2(kL_m) \tag{C.24}$$

–for the structure on Fig. 7.1b,

$$Y_{mn}^{R}(kL_m, kL_n) = \frac{4\pi}{a_R b_R} \sum_{m=1,3...}^{\infty} \sum_{n=0}^{\infty} \frac{\varepsilon_n(k^2 - k_x^2)}{kk_z} \coth k_z H$$
$$\times \cos k_y y_{0m} \cos k_y (y_{0n} + \frac{d_n}{4}) I_1(kL_m) I_1(kL_n), \tag{C.25}$$

$$Y_{m3}^{R}(kL_m, kL_3) = Y_{3m}^{R}(kL_3, kL_m) = \frac{4\pi}{a_R b_R} \sum_{m=1,3...}^{\infty} \sum_{n=0}^{\infty} \frac{\varepsilon_n}{k_z \mathrm{sh} k_z H}$$
$$\times \cos k_y y_{0m} \cos k_y (y_{03} + \frac{d_3}{4}) I_1(kL_m) I_2(kL_3), \tag{C.26}$$

$$m = 1, 2; \quad n = 1, 2,$$

$$Y_{33}^{R}(kL_3) = \frac{4\pi}{a_R b_R} \sum_{m=1,3...}^{\infty} \sum_{n=0}^{\infty} \frac{\varepsilon_n k}{k_z(k^2 - k_x^2)} \coth k_z H$$
$$\times \cos k_y y_{03} \cos k_y (y_{03} + \frac{d_3}{4}) I_2^2(kL_3), \tag{C.27}$$

$$Y_{33}^{Hs}(kL_3) = \mathrm{Si}4kL_3 - i\mathrm{Cin}4kL_3 - 2\cos kL_3$$
$$\times \left[2(\sin kL_3 - kL_3\cos kL_3)\left(\ln\frac{16L_3}{d_3} - \mathrm{Cin}2kL_3 - i\mathrm{Si}2kL_3\right) \right.$$
$$\left. + \sin 2kL_3 e^{-ikL_3} \right], \tag{C.28}$$

$$I_1(kL_m) = 2\left\{ \frac{k\sin kL_m \cos k_x L_m - k_x\cos kL_m \sin k_x L_m}{k^2 - k_x^2}\cos\frac{\pi L_m}{a} \right.$$
$$\left. - \frac{\left(\frac{\pi}{a}\right)\sin\frac{\pi L_m}{a}\cos k_x L_m - k_x\cos\frac{\pi L_m}{a}\sin k_x L_m}{(\pi/a)^2 - k_x^2}\cos kL_m \right\}, \tag{C.29}$$

$$I_2(kL_3) = 2\frac{k_x\sin kL_3\cos k_x L_3 - k\cos kL_3\sin k_x L_3}{k_x}. \tag{C.30}$$

In the (C.23)–(C.30) expressions there are the following symbols: $k_x = \dfrac{m\pi}{a\{a_R\}}$, $k_y = \dfrac{n\pi}{b\{b_R\}}$, $k_z = \sqrt{k_x^2 + k_y^2 - k^2}$, (m, n $= 0, 1, 2...$); $\varepsilon_n = 1$ at n $= 0$, $\varepsilon_n = 2$ at n $\neq 0$; y_{0m} are the slots axes coordinates; $Si(x)$ and $Cin(x)$ are the integral sine and cosine.

Appendix D
Series Summing Up in the Function of the Iris Own Field

The real part of the own field function $W\left(\frac{\pi d_e}{2L}, \frac{\pi}{2}\right)$ of the slot has been form:

$$\operatorname{Re}W\left(\frac{\pi d_e}{2L}, \frac{\pi}{2}\right) = \frac{4\pi^2}{abL} \left\{ \sum_{m=3,5...}^{\infty} \frac{1 + \cos 2k_x L}{k_z^{m0}[(\pi/(2L))^2 - k_x^2]} \right.$$

$$+ \frac{2\cos^2(\pi L/a)}{\gamma_{10}^2} \sum_{n=1}^{\infty} \frac{\cos(k_y d_e/4) + \cos 2k_y y_0}{\sqrt{k_y^2 - \gamma_{10}^2}} \quad \text{(D.1)}$$

$$\left. + \sum_{m=3,5...}^{\infty} \frac{1 + \cos 2k_x L}{(\pi/(2L))^2 - k_x^2} \sum_{n=1}^{\infty} \frac{\cos(k_y d_e/4) + \cos 2k_y y_0}{k_z^{mn}} \right\}.$$

The following symbols are accepted in (D.1): $k_z^{m0} = \sqrt{k_x^2 - (\pi/(2L))^2}$; $k_z^{mn} = \sqrt{k_x^2 + k_y^2 - (\pi/(2L))^2}$; $\gamma_{10}^2 = (\pi/(2L))^2 - (\pi/a)^2$; $k_x = m\pi/a$; $k_y = n\pi/b$.

The second series over n in (D.1) can be summed up with the help of the relation reduced in [5.6]:

$$\sum_{n=1}^{\infty} \frac{\cos(k_y d_e/4) + \cos 2k_y y_0}{k_z^{mn}} \cong -\frac{1}{k_z^{m0}} + \frac{b}{\pi}\left[K_0(k_z^{m0} \frac{d_e}{4}) + K_0(2k_z^{m0} y_0) \right], \quad \text{(D.2)}$$

where $K_0(x)$ is the McDonald's function. Then (D.1) grades into the expression that does not contain double series:

$$
\operatorname{Re} W\left(\frac{\pi d_e}{2L}, \frac{\pi}{2}\right) \cong \frac{4\pi^2}{abL}\left\{ \frac{2\cos^2(\pi L/a)}{\gamma_{10}^2} \sum_{n=1}^{\infty} \frac{\cos\dfrac{n\pi d_e}{4b} + \cos\dfrac{2n\pi y_0}{b}}{\sqrt{(n\pi/b)^2 - \gamma_{10}^2}} \right.
$$

$$
- \frac{b}{\pi}\left[\frac{2\cos^2\dfrac{3\pi L}{a}}{k_{30}^2}\left[K_0\left(k_{30}\frac{d_e}{4}\right) + K_0\left(2k_{30}y_0\right) \right] \right.
$$

$$
\left. \left. + \left(\frac{a}{\pi}\right)^2 \sum_{m=5}^{\infty} [1 - (-1)^m]\frac{1 + \cos\dfrac{2m\pi L}{a}}{m^2} K_0\left(\frac{m\pi d_e}{4a}\right) \right] \right\}. \tag{D.3}
$$

The term series (D.3) is separated out in the explicit form $(k_{30} = \sqrt{(3\pi/a)^2 - (\pi/(2L))^2})$ at $m = 3$ in (D.3), and in the rest sum along m we take into account that while it increases, the $K_0(x)$ function decreases rapidly.

Let us sum up the first series in (D.3), expanding the square root into series

$$
\left(\sqrt{(n\pi/b)^2 - \gamma_{10}^2}\right)^{-1} \approx \frac{b}{\pi}\left[\frac{1}{n} + \left(\frac{\gamma_{10}b}{\sqrt{2\pi}}\right)^2 \frac{1}{n^3}\right] \text{ and using the formula 1.441.2}
$$

from [5.10]: $\displaystyle\sum_{n=1}^{\infty} \frac{\cos nx}{n} = -\ln\left(2\sin\frac{x}{2}\right)$, valid under condition of $0 < x < 2\pi$, and

which in our case $(x = \dfrac{\pi d_e}{4b}$ and $x = \dfrac{2\pi y_0}{b})$ is performed. As result after some transformations we get:

$$
\sum_{n=1}^{\infty} \frac{\cos n\dfrac{\pi d_e}{4b} + \cos n\dfrac{2\pi y_0}{b}}{\sqrt{(n\pi/b)^2 - \gamma_{10}^2}} \cong \frac{2\cos^2\dfrac{\pi y_0}{b}}{k_{11}}
$$

$$
- \frac{b}{\pi}\left[2\cos^2\frac{\pi y_0}{b} - \left(\frac{\gamma_{10}b}{9}\right)^2 \cos^2\frac{2\pi y_0}{b} + \ln\left(\frac{\pi d_e}{2b}\sin\frac{\pi y_0}{b}\right) \right], \tag{D.4}
$$

where $k_{11} = \sqrt{(\pi/a)^2 + (\pi/b)^2 - (\pi/(2L))^2}$.

We can sum up the rest series along m in (D.3) with the help of the expressions 8.526.1,2 from [5.10]:

$$
\sum_{m=1}^{\infty} [1 - (-1)^m] K_0(mx) \cos mxt
$$

$$
\cong \frac{\pi}{2x\sqrt{1+t^2}} + \frac{1}{2}\left[\frac{1}{1 - (xt/(2\pi))^2} - \frac{2}{1 - (xt/\pi)^2} - \frac{2/3}{1 - (xt/(3\pi))^2} + \frac{1}{2} \right], \tag{D.5}
$$

which are valid while performing the following conditions: $x > 0$, $\{x/\pi\} \ll 1$, $\{xt/\pi\} < 1$ (here and further x and $t-$ are non-dimensional parameters,

m — integers). Making double integration along t in (D.5), we have:

$$\sum_{m=1}^{\infty} [1 - (-1)^m] K_0(mx) \frac{1 + \cos mxt}{m^2}$$

$$\cong -\frac{\pi x}{2} \left\{ t \ln(t + \sqrt{1 + t^2}) - \sqrt{1 + t^2} + 1 + t \ln \frac{\pi - xt}{\pi + xt} - \frac{\pi}{x} \ln \left[1 - \left(\frac{xt}{\pi} \right)^2 \right] \right.$$

$$- t \ln \frac{2\pi - xt}{2\pi + xt} + \frac{2\pi}{x} \ln \left[1 - \left(\frac{xt}{2\pi} \right)^2 \right] + t \ln \frac{3\pi - xt}{3\pi + xt} - \frac{3\pi}{x} \ln \left[1 - \left(\frac{xt}{3\pi} \right)^2 \right]$$

$$\left. + \frac{xt^2}{4\pi} \right\} + 2 \sum_{m=1}^{\infty} [1 - (-1)^m] \frac{K_0(mx)}{m^2}. \tag{D.6}$$

Then substituting the values $x = \frac{\pi d_e}{4a}$ ($\{d_e/(4a)\} \ll 1$), $t = \frac{8L}{d}$, $xt = \frac{2\pi L}{a}$ ($\{2L/a\} < 1$) in (D.6), we get:

$$\sum_{m=5}^{\infty} [1 - (-1)^m] \frac{1 + \cos m \frac{2\pi L}{a}}{m^2} K_0 \left(m \frac{\pi d_e}{4a} \right)$$

$$\cong \frac{\pi^2 L}{a} \left(1 - \ln \frac{16L}{d_e} + \ln \frac{1 - L/a}{1 + L/a} - \ln \frac{1 - 2L/a}{1 + 2L/a} - \ln \frac{1 - 2L/(3a)}{1 + 2L/(3a)} \right)$$

$$+ \frac{\pi^2}{2} \left\{ \ln \left[1 - \left(\frac{2L}{a} \right)^2 \right] - 2 \ln \left[1 - \left(\frac{L}{a} \right)^2 \right] + 3 \ln \left[1 - \left(\frac{2L}{3a} \right)^2 \right] - \left(\frac{L}{a} \right)^2 \right\}$$

$$+ 4 \left[K_0 \left(\frac{\pi d_e}{4a} \right) \sin^2 \frac{\pi L}{a} + \frac{1}{9} K_0 \left(\frac{3\pi d_e}{4a} \right) \sin^2 \frac{3\pi L}{a} + \sum_{m=5,7...}^{\infty} K_0 \left(m \frac{\pi d_e}{4a} \right) \bigg/ m^2 \right]. \tag{D.7}$$

The expressions (D.4) and (D.7) give the final formula (5.23) after substitution in (D.3).

Appendix E
Electromagnetic Values in CGS and SI Systems of Units

Because we use the CGS system in the book, it is expedient to make a brief comparison of main ratios of electromagnetic theory, represented in the CGS and SI systems.

Maxwell's equations in CGS system of units:

$$\text{rot } \vec{E} = -\frac{1}{c}\frac{\partial \vec{B}}{\partial t},$$

$$\text{rot } \vec{H} = \frac{1}{c}\frac{\partial \vec{D}}{\partial t} + \frac{4\pi}{c}\vec{j},$$

$$\text{div } \vec{D} = 4\pi\rho,$$

$$\text{div } \vec{B} = 0,$$

where the field sources are:

$\vec{j} = ne\vec{v}$ is the electric current density, n is the electron density, $\vec{v}$ is their velocity, e is the charge in CGS units;

$\rho = ne$ is the electric charge density.

Maxwell's equations in SI system of units:

$$\text{rot } \vec{E} = -\frac{\partial \vec{B}}{\partial t},$$

$$\text{rot } \vec{H} = \frac{\partial \vec{D}}{\partial t} + \vec{j},$$

$$\text{div } \vec{D} = \rho,$$

$$\text{div } \vec{B} = 0,$$

where the field sources are:

$\vec{j}$ is the electric current density in *ampere per square meter* (A/m^2);

ρ is the electric charge density in *coulombs per cubic meter* (C/m^3).

1 unit of electric charge in CGS system = $\frac{1}{3\cdot10^9}$ *coulombs*,

1 unit of electric current density in CGS system = $\frac{1}{3\cdot10^5}$ A/m^2.

Constitutive equations in CGS system:

$$\vec{D} = \varepsilon \vec{E}; \quad \vec{B} = \mu \vec{H}.$$

Constitutive equations in SI system:

$$\vec{D} = \varepsilon_0 \varepsilon \vec{E}; \quad \vec{B} = \mu_0 \mu \vec{H},$$

where the permittivity and permeability in the CGS system do not have dimensions and equal to the relative permittivity and permeability ε and μ in the SI system, ε_0 and μ_0 are the corresponding permittivity and permeability of vacuum here.

The electric and magnetic field intensities in the CGS system have the same dimension $g^{1/2} cm^{-1/2} sec^{-1}$, but intensity of the magnetic field is called *oersted* (Oe).

1 unit of electric field intensity in CGS system = $3 \cdot 10^4$V/m,
1 Oe = $\frac{1}{4\pi} \cdot 10^3$A/m.

The electric and magnetic inductions ($\vec{D}$ and $\vec{B}$) in the CGS system have dimensions of the intensities of corresponding fields. They have different dimensions and different names in the SI system.

In the SI system the electric induction is measured in *coulombs per square meter* (C/m^2), at this

1 unit of the electric induction in CGS system = $\frac{1}{12\pi} \cdot 10^{-5}$C/m^2,

and the magnetic induction amount is measured in *webers per square meter* = 1 *tesla* (T),

1 *gauss* (Gs) in CGS system = 10^{-4} T.

The potentials are defined nearly alike in both systems (the potentials of only electric type are represented here):

$$\vec{H} = \frac{1}{\mu} \operatorname{rot} \vec{A},$$

$$\vec{E} = -\operatorname{grad} \varphi - \frac{1}{c} \frac{\partial \vec{A}}{\partial t},$$

with Lorentz's condition $\frac{1}{c} \frac{\partial \varphi}{\partial t} + \operatorname{div} \vec{A} = 0$ in the CGS system and

$$\vec{B} = \operatorname{rot} \vec{A},$$

$$\vec{E} = -\operatorname{grad} \varphi - \frac{\partial \vec{A}}{\partial t},$$

with Lorentz's condition $\frac{1}{v^2} \frac{\partial \varphi}{\partial t} + \operatorname{div} \vec{A} = 0$ in SI system, where v is the phase velocity of the wave in the corresponding medium.

The wave equations also differ slightly:

$$\Delta\varphi - \frac{1}{v^2}\frac{\partial^2\varphi}{\partial t^2} = -\frac{4\pi\rho}{\varepsilon},$$

$$\Delta\vec{A} - \frac{1}{v^2}\frac{\partial^2\vec{A}}{\partial t^2} = -\frac{4\pi\mu}{c}\vec{j}$$

in the CGS system;

$$\Delta\varphi - \frac{1}{v^2}\frac{\partial^2\varphi}{\partial t^2} = -\frac{\rho}{\varepsilon_0\varepsilon},$$

$$\Delta\vec{A} - \frac{1}{v^2}\frac{\partial^2\vec{A}}{\partial t^2} = -\mu_0\mu\,\vec{j}$$

in the SI system, and the phase velocity of the plane wave equals

$$v = \frac{c}{\sqrt{\varepsilon\mu}}$$

in the first case and

$$v = \frac{1}{\sqrt{\varepsilon_0\mu_0}}\frac{1}{\sqrt{\varepsilon\mu}} = \frac{3\cdot 10^8\,[\text{m/sec}]}{\sqrt{\varepsilon\mu}}$$

in the second case because

$$\mu_0 = 4\pi\cdot 10^{-7}\frac{\text{H}}{\text{m}}, \quad \varepsilon_0 = \frac{1}{36\pi}10^{-9}\frac{\text{F}}{\text{m}},$$

and *henry·farad* = *sec*^{2}. Thus the finite result is the same.

The wave equations solutions can be proposed in the form of general potentials in both cases, and they are written with the help of the Fourier components of these potentials in the form of:

$$\vec{E}(\vec{r}) = \left(\text{grad div} + k^2\varepsilon\mu\right)\vec{\Pi}(\vec{r}),$$

$$\vec{H}(\vec{r}) = \frac{ik}{w}\,\text{rot}\vec{\Pi}(\vec{r}),$$

where in the CGS system

$$k^2 = \frac{\omega^2}{c^2}, \quad w = 1, \quad \vec{\Pi}(\vec{r}) = \frac{1}{i\omega}\int_V \frac{\vec{j}(\vec{r}')}{|\vec{r}-\vec{r}'|}e^{-ik|\vec{r}-\vec{r}'|}d\vec{r}',$$

in the SI system

$$k = \omega\sqrt{\varepsilon_0\mu_0}, \quad w = \sqrt{\frac{\mu_0}{\varepsilon_0}} \text{ is the wave impedance of free space,}$$

$$\vec{\Pi}(\vec{r}) = \frac{1}{4\pi i \omega \varepsilon_0} \int\limits_V \frac{\vec{j}(\vec{r}')}{|\vec{r} - \vec{r}'|} e^{-ik|\vec{r} - \vec{r}'|} d\vec{r}'.$$

Thus, the electromagnetic wave in free space in the SGS system is designated as mutually orthogonal triplet of the vectors $\vec{E}$, $\vec{H}$ and $\vec{k}$ where $\vec{E}$ and $\vec{H}$ have equal dimension and value, and this very triplet consists already of values that have different names and dimensions—*volt/meter*, *ampere/meter* and the wave vector with dimension 1/*meter* sin the SI system. But the wave has a simple physical meaning—a long line in this case.

Index

Printed in the United States of America

RETURN TO: PHYSICS-ASTRONOMY LIBRARY
351 LeConte Hall   510-642-3122

LOAN PERIOD 1 **1-MONTH**	2	3
4	5	6

ALL BOOKS MAY BE RECALLED AFTER 7 DAYS.
Renewable by telephone.

DUE AS STAMPED BELOW.

FORM NO. DD 22
2M 7-07

UNIVERSITY OF CALIFORNIA, BERKELEY
Berkeley, California 94720–6000